Changing Landscapes

This book is dedicated to John Wyatt-Smith, who pioneered the conservation management of tropical rain forests

Changing Landscapes

The Development of the International Tropical Timber Organization and its Influence on Tropical Forest Management

Duncan Poore

ITTO

EARTHSCAN

Earthscan Publications Ltd
London • Sterling, VA

First published in the UK and USA in 2003
by Earthscan Publications Ltd

ISBN: 1-85383-991-4 (paperback)
      1-85383-990-6 (hardback)

Typesetting by MapSet Ltd, Gateshead, UK
Printed and bound in the UK by Creative Print and Design (Wales), Ebbw Vale
Cover design by Danny Gillespie

For a full list of publications please contact:

Earthscan Publications Ltd
120 Pentonville Road, London, N1 9JN, UK
Tel: +44 (0)20 7278 0433
Fax: +44 (0)20 7278 1142
Email: earthinfo@earthscan.co.uk
Web: **www.earthscan.co.uk**

22883 Quicksilver Drive, Sterling, VA 20166-2012, USA

A catalogue record for this book is available from the British Library

Library of Congress Cataloging-in-Publication Data

Poore, Duncan.
  Changing landscapes : the development of the International Tropical Timber
Organization and its influence on tropical forest management / Duncan Poore.
      p. cm.
Includes bibliographical references and index (p.).
  ISBN 1-85383-990-6 (hardback) – ISBN 1-85383-991-4 (paperback)
  1. Forest management–Tropics. 2. Sustainable forestry–Tropics. 3. International
Tropical Timber Organization. I. Title.

  SD247.P64 2003
  634.9'0913–dc21

                                                                    2003003706

Earthscan is an editorially independent subsidiary of Kogan Page Ltd and publishes in
association with WWF-UK and the International Institute for Environment and
Development

This book is printed on elemental chlorine-free paper

Contents

List of figures, tables and boxes

Figures

Tables

Boxes

Foreword

One of the most contentious issues at the World Summit on Sustainable Development in Johannesburg in 2002 was the effectiveness of the various international environmental instruments and processes. Non-governmental organizations (NGOs) have blamed governments for not delivering on the commitments they have made in global conventions. Governments respond that the commitments are so general that it is difficult to translate them into practical action on the ground. This has led several governments to favour the so-called Type 2 initiatives to tackle specific problems – partnerships amongst concerned governments, civil society and the private sector.[1] The NGOs have countered that this is just an excuse to avoid the really difficult issues.

Two instruments that have been much criticized for having generated many words but not much action are the Convention on Biological Diversity (CBD) and the Intergovernmental Panel, and the subsequent United Nations Forum, on Forests.[2] Both these instruments had very broad mandates and they came into existence six years after the International Tropical Timber Organization (ITTO). It is therefore interesting to compare the performance of these three bodies. ITTO has international treaty status, but otherwise it has some of the characteristics of the Type 2 partnerships that were the subject of so much attention at the Johannesburg summit. ITTO was established in response to a specific problem – to ensure that internationally traded tropical timber came from sustainably managed forests. State parties were the countries that produced the timber and those that consumed it. The private sector timber industry and conservation NGOs have always been influential at meetings of ITTO's Council. All the actors concerned with the problem were thus part of the process. ITTO was also set up in a way that allowed it to intervene at all levels, from intergovernmental and national policy levels through to field projects. It was a forum with policy influence and a capacity to test and demonstrate policies and technologies on the ground. It is the only international environmental instrument with this on-the-ground project capacity.

Readers of this book will form their own opinions about the impact of ITTO relative to the other instruments mentioned. My own view is that the combination within ITTO of a multi-stakeholder negotiating framework and a mechanism for the delivery of projects to solve problems on the ground has been unique and successful. But a large part of its success has also come from its ability to attract the leading thinkers on and practitioners of sustainable forestry to its ranks. Duncan Poore was one of these. He was there even before the Organization was established and he brought to it his experience of practical field ecology in one of the most important tropical forest countries, Malaysia.

Later he became Director General of the Nature Conservancy of Great Britain – a distinguished body operating practical conservation programmes based on hard scientific analysis. He served as scientific director and later as chief executive of the International Union for the Conservation of Nature and Natural Resources (IUCN), itself a unique union of governmental and non-governmental organizations. Later he was director of the Commonwealth Forestry Institute at Oxford – also a unique blend of hard science and an extensive network of practical foresters.

It is sometimes said that tropical forests are a nightmare for foresters but a paradise for ecologists. It is perhaps symptomatic of this that Duncan Poore is an ecologist. The complexity and diversity that characterize tropical forests fascinate ecologists. Several ecologists, including Poore, have been led to a deep interest in and empathy for the people who inhabit, use and have a treasure house of knowledge stored in these forests. The ecological and human dimensions of tropical forests have been prominent in the work of ITTO and this is in no small measure due to Poore's efforts.

Poore has never held a staff position at ITTO. Rather he has been its *éminence grise*. He led landmark studies, one of which produced the seminal book *No Timber without Trees* (Poore et al, 1989), a publication that had profound impacts on tropical forestry and on ITTO. He was the inspiration behind the ITTO guidelines series – a concept that he had first developed with his *Ecological Guidelines for Development in Tropical Rain Forests* (Poore, 1976) produced for IUCN's Commission on Ecology in the 1970s. He also took part in the influential and challenging mission to assess the sustainability of forestry in the Malaysian state of Sarawak in 1989–1990 – a mission that set the pattern for a number of subsequent studies in other countries.

Duncan Poore has been a powerful intellectual force behind ITTO and it is fitting that he should be the person to attempt to synthesize and capture the history of the organization in this volume. In doing so, he has drawn on his unique experience of ITTO and of the evolution of international actions dealing with tropical forests in general.

So what are the achievements of ITTO? It has certainly enriched our understanding of tropical forest issues – the variety of perspectives it has brought to the debate has been unique and valuable. The ITTO guidelines have influenced national policies and laws and have been put into practice by forest operators in many countries and situations – although perhaps not as faithfully nor on the scale that one might have wished. The criteria and indicators for sustainable management have been widely used and have certainly led to improved practice – again, perhaps not on the scale that might have been hoped for. The target for the achievement of sustainable sources of all traded tropical timber created pressure and incentives for better management in many countries. Missions to 'problem' countries have raised the political profile of forest issues and have led to political and practical change in a number of situations. The projects have a mixed record, but one can point to a number of significant successes, especially in promoting conservation landscapes where protected areas and sustainable forestry have been developed to complement one another. Added to this is a rich history of workshops and technical meetings that have helped us all to learn and develop professionally.

Thus, ITTO can legitimately claim to have been a power for good in achieving the sustainable management of tropical forests – and the influence of Duncan Poore in achieving this has been considerable. Whether it is legitimate to try and compare the impact of ITTO with that of other conventions and fora is a moot point. Those others set out to do different things. Perhaps it is best to regard them all as a nested hierarchy of interventions that have differing but complementary roles. Perhaps the problem that has surfaced in Johannesburg is that we have an imbalance between the general global debates and the more focused problem-solving instruments. ITTO has demonstrated the potential of the latter and perhaps we need more of these sorts of targeted partnerships.

This book is also looking to the future. The problems of tropical forests are by no means solved. The demands for forest products and services are growing and pressures on the land are increasing. Consumers have to be convinced that their timber is from sustainable sources; producers have to make more money out of timber than they could out of alternative land use. ITTO is well placed to help achieve these two goals and Duncan Poore's book shows us the way forward.

Jeffrey A Sayer
Professor and Senior Associate, World Wide Fund For Nature
International
Charge de Mission, CIRAD-Forêt, Montpellier, France
(Director General CIFOR, 1992–2001)

Preface

This book is mainly about the International Tropical Timber Organization (ITTO). It describes the origins of ITTO, its aspirations, the development of its policies and programmes, the interplay of political interests that has shaped it, its strengths and weaknesses, and its successes and failures. ITTO was set up in the confident belief that a flourishing timber trade would promote the wise management of tropical forests and thus the perpetuation of the resource. I explore how far this belief has been translated into reality. At the same time, views about what constitutes wise (sustainable) management have been changing; I discuss the significance of these changes and the difficulties, confusions and opportunities that they have created.

But the book also addresses a broader theme in its first and last chapters – the historical changes that have taken place in tropical forests and the wider influences that affect these forests. It ends by speculating about their possible future. Throughout their history, forests have been the playthings of powerful external forces besides those of trade – including changes of climate, population, agriculture, sources of energy, war and many others. Any realistic planning for the future of forests cannot ignore these factors; forests must not be treated in isolation.

I have attempted to make my account of the history of ITTO as objective as possible, based mainly on records in the ITTO archives; it should be obvious where personal views are inserted. But no history can be wholly objective for it depends on a selection from the available evidence and even the records are selective. My choice has been in favour of material that has a bearing on forest management. As a result, there are many aspects of the work of ITTO that have not been included, such as the development of market issues, so fully covered in the records of the annual market discussions – a subject for another book.

I have also been influenced by my own experience, which has spanned the whole period in which interest and concern about tropical forests has developed. This is an opportunity to pay tribute to many people who have shaped my thinking over the years and to those who have helped me in the writing of this book. Prominent among them have been the following: Sir Harry Godwin and AS Watt, who taught brilliantly in the late 1940s about the dynamics of vegetation and the influence of climate change; John Wyatt-Smith, for inspiring me with an interest in the Malaysian forests and demonstrating that they could, with skill and dedication, be managed for timber and still retain their other values; John Corner, Peter Ashton, Tim Whitmore, Gathorne Cranbrook and Frank Wadsworth, for their insights into tropical forest ecology; Michel Batisse and his team in UNESCO, for the privilege of working with them in developing the concept of the Biosphere Reserve; and Max Nicholson, Ray Dasmann, Gerardo Budowski, Sir Martin Holdgate and Jeff Sayer, who have been colleagues in conservation and research; Brian Johnson, Mike Ross, Ronald van der Giessen, Simon Rietbergen, Julian Evans, Caroline Sargent and Steve Bass, for the enthusiasm, drive and imagination they brought to the Forestry and Land Use Programme of IIED; Alf Leslie, Terence H'pay and Katsuhiro Kotari, prominent among the intellectual fathers of ITTO; Dr Freezailah bin Che Yeom, the first executive director, for involving me so deeply in fascinating work for the Organization; and Geoffrey Pleydell and Arthur Morrell, who guided me on trade issues. Arthur tragically died while attending an ITTO Council session.

This book was commissioned by ITTO. It would not have been possible without the unfailing encouragement of Dr Manoel Sobral Filho, the present executive director of ITTO, and his staff – particularly Alastair Sarre and Mike Adams – and the wonderful logistical support of their secretaries. I have also developed personal friendships with many delegates to meetings of the Council of ITTO and have benefited greatly from discussions and arguments with them. Thang Hooi Chiew has been a wonderful and stimulating colleague in much of the recent work I have done for ITTO, including the review of progress towards the Year 2000 Objective, which was the basis for Chapter 14. It would have been impossible to write Chapter 3 without the help and recollections of Terence H'pay, Alf Leslie and Geoffrey Pleydell. The text has been much improved by comments and material from many of the above, and from Jim Ball, Jim Bourke, David Kaimowitz, Jagmohan Maini, Douglas Pattie, Catriona Prebble and Peter Wood. I am most grateful to all.

My wife, Judy, has been an unfailing support and vigorous critic.

Duncan Poore
Glenmoriston
September 2002

List of acronyms and abbreviations

AAC	annual allowable cut
AIMA	Asociación Ecuatoriana de Industriales de la Madera
APKINDO	Indonesian Panel Products Manufacturers Association
ASEAN	Association of Southeast Asian Nations
ATIBT	Association Technique International des Bois Tropicaux
ATO	African Timber Organization
BP	before present ('present' being taken as 1950 AD)
CAR	Central African Republic
CBD	Convention on Biological Diversity
CFA	Commonwealth Forestry Association
CFA	Currency of Francophone Africa
CGIAR	Consultative Group on International Agricultural Research
CIDA	Canadian International Development Agency
CIFOR	Center for International Forestry Research
CITES	Convention on International Trade in Endangered Species of Wild Fauna and Flora
CODECHOCO	Choco Sustainable Development Corporation
COICA	Coordinadora de las Organizaciónes Indigenas de la Cuenca Amazónica
CPF	Collaborative Partnership on Forests
CSCE	Conference on Security and Cooperation in Europe
CSD	Commission on Sustainable Development
dbh	diameter at breast height
EC	European Community
ECE	Economic Commission for Europe (United Nations)
ECOSOC	Economic and Social Council (United Nations)
EIA	environmental impact assessment
EU	European Union
FAO	Food and Agriculture Organization (United Nations)
FFDC	Forestry Forum for Developing Countries
FOB	free on board (value includes only the cost of the freight)
FoE	Friends of the Earth
FRA	Forest Resource Accounting
FRA	Forest Resource Assessment
FRIM	Forest Research Institute Malaysia
FSC	Forest Stewardship Council
GATT	General Agreement on Tariffs and Trade

GDP	gross domestic product
GIS	geographical information system
GLOBE	Global Legislators Organisation for a Balanced Environment
GOI	Government of Indonesia
GNP	gross national product
GSP	generalized scheme of preferences
GTZ	Organization for Technical Cooperation (Germany)
HIID	Harvard Institute of International Development
HPH	Hak Pengusahaan Hutan (logging concession rights – Indonesia)
IBRD	International Bank for Reconstruction and Development
IAF	International Arrangement on Forests
IFF	Intergovernmental Forum on Forests
IHPA	International Wood Products Association
IIED	International Institute for Environment and Development
IISD	International Institute for Sustainable Development
ILO	International Labour Organization
IMF	International Monetary Fund
INRENA	Peruvian National Institute for Natural Resources
IPC	Integrated Programme for Commodities
IPF	Intergovernmental Panel on Forests
IPCC	Intergovernmental Panel on Climate Change
ISO	International Organization for Standardization
ITC	International Trade Centre
ITTA	International Tropical Timber Agreement
ITTC	International Tropical Timber Council
ITTO	International Tropical Timber Organization
IUCN	International Union for Conservation of Nature and Natural Resources (World Conservation Union)
JLIA	Japanese Lumber Importers' Association
JOFCA	Japanese Overseas Forestry Consultants Association
LEEC	London Environmental Economics Centre
LEI	Indonesian Ecolabelling Institute
MAB	Man and the Biosphere
MAI	mean annual increment
MCPFE	Ministerial Conference for the Protection of Forests in Europe
MDF	medium density fibreboard
MIS	Tropical Timber Market Information Service
MOEF	Ministry of Environment and Forests (Government of India)
NCR	Native Customary Rights (Sarawak)
NFAP	National Forests Action Programme
NFATT	Netherlands Framework Agreement on Tropical Timber

NFP	National Forest Programme
NGO	non-governmental organization
NWFP	non-wood forest product
ODA	official development assistance
ODA	Overseas Development Administration (UK)
OECD	Organisation for Economic Co-operation and Development
ONADEF	National Forest Development Agency (Cameroon)
PEFC	Pan-European Forest Certification
PFE	permanent forest estate
PFMA	sustained forest management plan (Brazil)
PNG	Papua New Guinea
RIL	reduced impact logging
SAM	Sahabat Alam Malaysia
SERNAP	Bolivian National Service for Protected Areas
SFM	sustainable forest management
STA	Sarawak Timber Association
STIDC	Sarawak Timber Industry Development Corporation
TFAP	Tropical Forestry Action Plan (later Tropical Forests Action Programme)
TFAP-CG	TFAP Consultative Group
TFF	Tropical Forest Foundation
TPA	totally protected area
UCBT	Union Commerciale des Bois Tropicaux
UN	United Nations
UN COMTRADE	United Nations Commodity Trade Flow (database)
UNCED	United Nations Conference on Environment and Development
UNCTAD	United Nations Conference on Trade and Development
UNDP	United Nations Development Programme
UNEP	United Nations Environment Programme
UNESCO	United Nations Educational, Scientific and Cultural Organization
UNFCCC	United Nations Framework Convention on Climate Change
UNFF	United Nations Forum on Forests
UNIDO	United Nations Industrial Development Organization
VJR	Virgin Jungle Reserve (Malaysia)
WCMC	World Conservation Monitoring Centre
WCSD	World Commission on Sustainable Development
WRI	World Resources Institute
WHO	World Health Organization
WWF	World Wide Fund For Nature

1

The rise and fall of forests

'Although some societies may have lived, and perhaps some still live, in ecological harmony with their surroundings, depending solely on local resources and using these sustainably, this seems to have been a rare occurrence'

It may seem rather perverse to begin a book about forests and trade in this century with a flashback to the descent of the precursors of our species from the trees some 4 million years ago; but it is far from irrelevant. From that time onwards, the future of forests and of human beings have been intertwined – for land, food, shelter, fuel, medicine, materials and as a subject for belief in the sacred and supernatural.

Closed-canopy tropical rain forest is thought to have first become widespread in the Palaeocene (54–66 million years ago), when the climate was warmer than today and the continents were not yet entirely in their present positions. The forest then was more widespread than now, occurring not only in belts on both sides of the equator but also in a discontinuous band outside the tropics in all four continents. The Middle Miocene (10–16 million years ago) was again a period of warm climate – the continents having assumed more or less their present positions. The distribution of the forest in the Middle Miocene is shown in Figure 1.1. After that, the climate deteriorated, the forest withdrew to the tropical zone, and grasslands and deserts expanded.[1]

In Equatorial Africa, as the dense forest retreated towards the west, the first separation took place between the ancestral line of the apes, who remained in the forest, and the ancestors of man,[2] who took to the plains and lived consecutively in open woodland, arid scrub-land, and savannas and open grasslands. Figure 1.2 shows the differentiation of hominid species in Africa with an indication of the habitats they occupied and the food they preferred.

Early *Homo sapiens* lived and moved in open country – through the belts of savanna and steppe that lay between the tropical forests and the temperate forests to the north and south. By the time that these great movements of people were taking place, the world was in the grip of the Pleistocene glaciations. The distribution of forests then was very different from that of today; the present temperate and boreal regions were covered with ice and the rain forests of the tropics were much restricted (see Figures 1.3, 1.4 and 1.5).

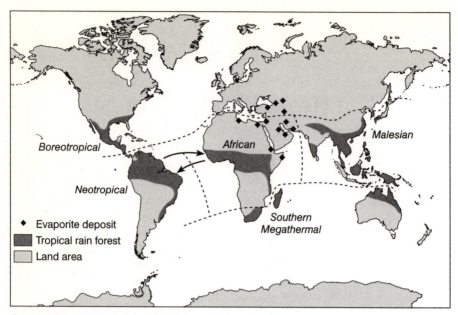

Figure 1.1 *Distribution of closed-canopy tropical rain forests during the Middle Miocene, coinciding with the Miocene thermal maximum*

Early men and women were very much a part of nature, subject to all the vicissitudes of heat, cold, droughts, floods, lightning, fire and natural disasters. Modern man is, of course, just as much a part of nature as his ancestors; it is only that, for a long time, the fortunate have been able to insulate themselves from direct exposure to elemental forces. It may be, however, that the insulation is now wearing rather thin. It is mistaken – and even dangerously misleading – to talk of man and 'the' environment. We are all part of nature; you are part of my environment just as much as I am part of yours; and everything that we do affects our surroundings in some way, however trivial.

The climate of the Earth has always changed and so have the configurations of the continents. The main belts of vegetation – equatorial and tropical forests, sub-tropical forest, savanna and steppe, semi-desert and desert, and temperate and boreal forests – have ebbed and flowed to settle in a dynamic equilibrium with the prevalent climate and distribution of land masses and oceans. In the case of forests, expansion would have meant the slow advance of the forest edge into grassland or heath as conditions gradually became suitable for the establishment of seedlings and young trees outside the forest – probably a scatter of trees rather than a solid front. Some young trees might leap considerable distances if they found a congenial niche. On the retreating edge, there would be death and failure of regeneration, leading to replacement of forest by other plant communities. In every instance, one must envisage these processes in terms of the germination, establishment, growth, fruiting and survival of individual trees: if conditions are suitable, advance; if unsuitable, retreat.

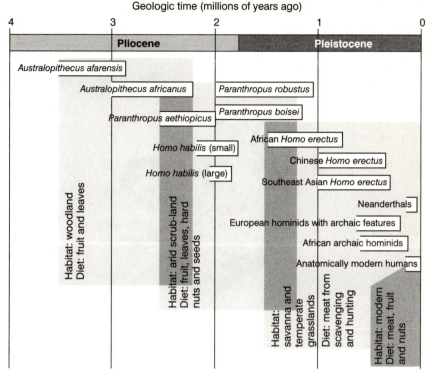

Geologic time (millions of years ago)

Source: Stringer and McKie, 1996.

Figure 1.2 *Fossil hominids through time, showing probable habitat and dietary changes*

The early history of man involved gradual migration – starting about 100,000 years ago, from his presumed origins in East Africa into Europe, the Middle East, Central Asia and China and thence across the land bridge into North America and across the seas into Australia and Polynesia. The probable course and timing of these movements is shown in Figure 1.6. Routes followed the lines of least resistance, along the coasts or by the corridors between the forest and the desert, or between the forests and the ice sheets. These were hunter-gatherers – people who lived off the fish and wild animals they could hunt or trap, and by the roots, leaves and fruits they could collect. They avoided deep forest and kept to the steppes, savannas and drier woodlands where, no doubt, they collected wood to construct shelters and fires for cooking and used fire to flush out game. We do not really know how great the direct effect of this way of life may have been on the vegetation they passed through, but it is likely to have been relatively slight. However, there is strong evidence that the story was very different for the fauna, and this would certainly have had some repercussions on the vegetation.

The appearance of human beings in the Americas and Australia was followed by the disappearance of many of the large animals that were there when they arrived.[3] Later, the same happened in New Zealand, which was

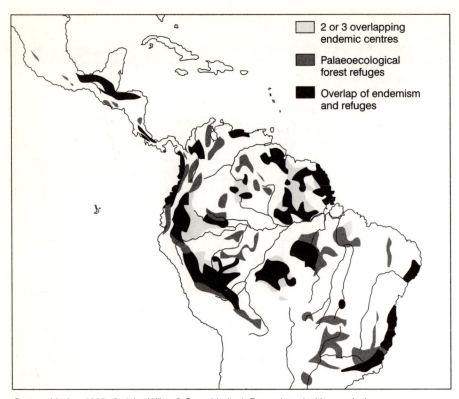

Figure 1.3 *Pleistocene refugia in New World rain forests*

colonized in the 11th century by people from Polynesia. It is particularly intriguing that extinction to the same degree does not seem to have happened in Africa or Eurasia. It is true that some large mammals such as the mammoth and the auroch did disappear, but generally most species seem to have survived. The hypothesis is that in Africa and Eurasia – but not in the other continents – the mammals and man evolved in constant contact with one another and that, accordingly, at the same time as man devised new methods of hunting, the mammals evolved new ways of avoiding capture or death. In the other continents, having evolved in the absence of man, they were unprepared and succumbed.

During this phase of human evolution, life was entirely based on the use of natural resources; life expectancy was short, most people apparently died of injuries, and human beings were still thin on the ground. We do not know whether there was any conscious effort towards living within the limits of local resources – what we might now call sustainable living. It is likely that people followed their prey, and when one site was hunted out, they moved on.

From the earliest times, there has been trade. Indeed it seems that the urge to barter and trade lies deep in human nature and, with developed speech, is one of the characteristics that separates mankind from other animals. Traded

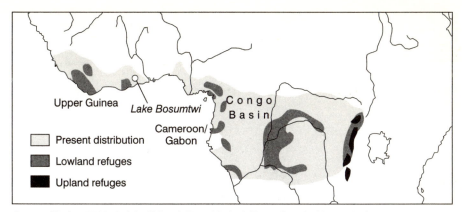

Source: Morley, 2000. © John Wiley & Sons Limited. Reproduced with permission.

Figure 1.4 *Rain forest refugia in Equatorial Africa in the last glacial maximum (c 18,000 years BP)*

obsidian from Turkey is present in the Pre-pottery Neolithic site of Beitha in the Jordan Valley dating from between 8000 and 6500 BC; and, of course, the development of the Bronze Age depended entirely on trade in both copper and tin. Much of the economic life of the ancient world depended upon trade.

It is difficult to be sure when trade in timber started, but it clearly had become important by the early historical period. Timber for shipbuilding in Athens came from Macedonia and the Black Sea and even from Sicily.[4] There was extensive trade in timber in the Mediterranean during the classical period, the cedars of Phoenicea being of special value. Alexander the Great arranged for sissoo wood (*Dalbergia sissoo*) to be imported from the Punjab for the pillars of his palace at Susa.[5] There are records that timber was shipped from Labrador to Iceland between the end of the 10th century and the middle of the 14th century.[6] But the attraction of wood did not necessarily lie only in the timber. The most important enticement to the Portuguese in Brazil in the 16th century was the dye *brasile* (from which the country took its name) extracted from the wood of the *pau do brasil* or brazilwood (*Caesalpinia echinata*). A flourishing and profitable trade was established with the local Indians in exchange for metal tools. And the spice trade was the great stimulus to the early European sea voyages to the East.

This begins to set the scene for this book – the interplay between the driving force of advancing technology, man's exploitation of natural resources and his propensity for trade.

The pressure of an increasing population eventually forced most human societies to abandon their hunter-gatherer way of life and to become settled farmers and graziers. The crops and stock upon which they based their farming differed from continent to continent and was dependent on the presence of wild species that could be domesticated. Where there were no suitable wild species, no farming appeared (see Figure 1.7). As populations grew, cultivation gradually extended to take in most cultivable soils in any region possessing a

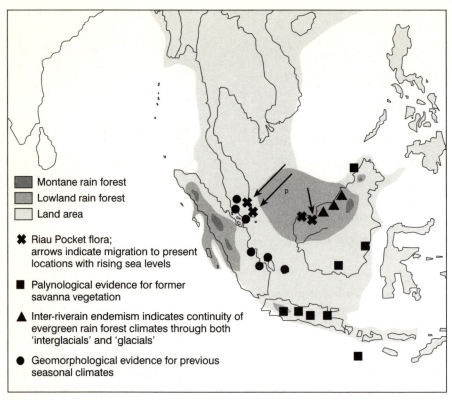

Montane rain forest

Lowland rain forest

Land area

✖ Riau Pocket flora;
arrows indicate migration to present
locations with rising sea levels

■ Palynological evidence for former
savanna vegetation

▲ Inter-riverain endemism indicates continuity of
evergreen rain forest climates through both
'interglacials' and 'glacials'

● Geomorphological evidence for previous
seasonal climates

Source: Morley, 2000. © John Wiley & Sons Limited. Reproduced with permission.

Figure 1.5 *Southeast Asian climate and vegetation (c 18,000 years BP)*

suitable climate for agriculture – first those that were light and easy to cultivate, and later, as better tools were devised, the heavier and more recalcitrant soils. Thus, by the end of the 19th century, cultivation and settlement had spread almost universally throughout all the parts of the world where settlement was not limited by climate (either too cold or too dry), by infertile soils, or by the prevalence of disease. Such expansion was all at the expense of natural vegetation, either forest or grassland.

This, of course, is an over-simplified picture. Hunter-gatherers still remained, at low densities, in some places. Nomadic graziers continued on the fringes of the deserts and the Arctic. Many communities depended upon both farming and hunting or fishing, and many agricultural areas retained portions of forest.

Although some societies may have lived, and perhaps some still live, in ecological harmony with their surroundings, depending solely on local resources and using these sustainably, this seems to have been a rare occurrence. More usually, where the soil was robust and fertile the population multiplied and overflowed into denser settlements (especially where irrigation was possible); where the soil was fragile, populations became less capable of supporting their

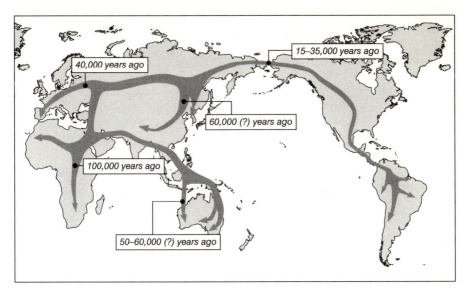

Source: Stringer and McKie, 1996.

Figure 1.6 *The spread of* Homo sapiens *over the last 100,000 years*

original numbers. Sophisticated societies were already, at this stage in their development, creating 'ecological footprints'. For example, the arid limestone hills around Athens were, by the 5th century BC, becoming incapable of supporting the urban population, and colonies were established on the north shores of the Black Sea to provide supplies of grain and fish.[7] Rome, too, depended on imports for its supplies of cereal – in this case from North Africa.

In fact, a dichotomy gradually developed, with the better soils being carefully farmed and husbanded with a liberal input of labour, manure and, if needed, irrigation water, while the remaining accessible natural ecosystems were subjected to extractive uses. There was a cumulative increase of yield from the former, whereas the latter might be used sustainably or not, according to circumstances.

By the beginning of the 20th century, farming for food had increased greatly, keeping pace with the growth of populations. But land was also being cleared on a large scale for commercial crops for export – particularly sugar cane and cotton, shortly to be followed by, among others, rubber, oil palm, cocoa, coffee and bananas. The only great expanses of almost undisturbed, pristine forest were in the boreal and equatorial regions, where farming was difficult or impossible and living conditions unpleasant. Even in the humid tropics, much forest had been cleared, especially in seasonally dry areas endowed with rich alluvial or volcanic soils; such were the conditions, for example, in the fertile plains of Java, Bali and Thailand. It has been estimated that, by this time, the world's forests had been reduced to about half their original area.[8] Apart from the boreal and equatorial forests, those that remained were: forests in climates suitable for agriculture which had not been cleared because they were

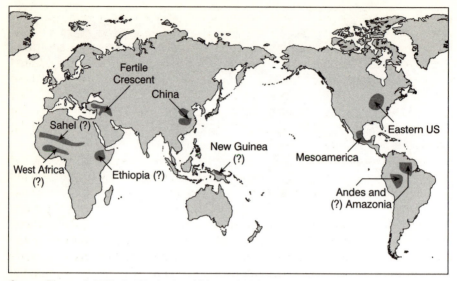

Figure 1.7 *Centres of origin of food production*

on poor or difficult soils or in areas where there was not yet a demand for more farmland; forests in populated areas which were retained because their wood was valued locally; and hunting reserves. But, in contrast, a number of areas had reverted to forest because a civilization had collapsed (such as the mahogany-rich forests growing on Mayan sites in central America) or because the people who had cleared them had moved on (such as the secondary hardwood forests of the eastern US).

Reversion to forest is still continuing, impelled by changes in socio-economic conditions. As traditional shepherding becomes uneconomic and socially unacceptable, traditional grazing lands in many parts of the Mediterranean basin, such as in Cyprus, Crete and Mediterranean France, are once again becoming covered with trees. In fact, in the course of human history, vast areas have fluctuated between forest and non-forest.

Another great surge of change came with the advent of mechanization. Suddenly there were many new opportunities for land-clearing and dam-building: communications improved dramatically, cold storage enabled the long-distance transport of perishable commodities and, at the same time, populations burgeoned and wealth and consumption increased.

This trend took off after the Second World War. There was a period in which land use surveys were carried out almost everywhere, not least in the tropics. These were mainly directed towards the expansion of agriculture, the inauguration of new irrigation schemes and the improvement of pasture. The fashionable land classification schemes of the time advocated a hierarchy of uses. If land was suitable for irrigated agriculture, it should be allocated to that use; next came dry-land agriculture; next, pasture. If land was unsuitable for any of these uses it was allocated to commercial forestry; last of all, the residue was considered suitable for the conservation of nature. Much land was converted to agriculture for food or commercial crops and many ambitious irrigation schemes were initiated; efforts were also made to improve the productivity of grazing lands. Forests were cleared wholesale for what were seen as more lucrative or necessary uses. In all these schemes, the overwhelming objective was greater productivity. The maintenance of fertility was, therefore, an important objective, and to that extent productive land was usually managed 'sustainably'; but this sustainability now involved the continued use of chemical fertilizers and large inputs of mechanical energy.

Perhaps because of the much longer time involved between harvests, the idea of sustained yield first appeared in forestry. A crop was harvested from the natural forest and this, in due course, was replaced by new growth that could then be harvested again. Systems of this kind had been applied for many centuries, with greater or lesser consistency, in medieval Europe and elsewhere. They were also applied in some parts of the tropics; sustained yield forestry was introduced to Burma, for example, in the 19th century; and efforts were made in both the British and French colonies to develop schemes of sustainable silviculture in the first half of the 20th century.

However, these tentative efforts were overwhelmed by the expansion of forest exploitation that came with mechanization. The advent of the chainsaw and extraction by tractors ushered in a phase of large-scale forest clearance for agriculture and of mining for timber. Large-scale land development schemes, often associated with huge water impoundments, became all the rage; some were well planned and successful (the development of rubber and oil palm in Peninsular Malaysia), others were disastrous (the East African groundnuts scheme and many of the Indonesian transmigration schemes). The pace of timber extraction far outstripped any attempts to manage for sustained yield, especially in East Asia where the forests were rich in economic species and there were ready markets.

The trends in Africa, South America and the Caribbean were broadly similar, but differed in detail. Neither continent had forests that contained such high volumes of merchantable timber, so the damage caused in extraction was not so great. At this time, many tropical countries also began to establish plantations of fast-growing species, especially *Eucalyptus* and *Pinus*. These were considered simpler to manage and had higher growth rates than the species of the indigenous forest. The argument that was increasingly put forward was that plantations would reduce pressure on the natural forest.

This, broadly, was the position when the International Tropical Timber Agreement (ITTA) came into force. The next chapter examines how the *idea* of sustainable forest management developed from the 1960s onwards. Chapters 3 to 17 concentrate on the history of the International Tropical Timber Organization (ITTO) and the contribution that it has made to furthering the concept and helping make it a reality.

The last chapter looks at future prospects. Suffice it to say now that: there is still pressure for farmland in many countries; mining for timber is still prevalent; the area still covered by untouched primary forest or old growth forest is rapidly declining; and there are large areas of secondary or degraded forest (and deforested land) which are amenable to restoration. Equally there are areas that, if subjected to further misuse, may become degraded beyond repair.

As we shall see, three main forces lay behind the creation of ITTO: on the part of tropical timber producing countries, there was the desire to maximize the contribution of forests to national development; on the part of tropical timber importing countries, there was the desire for a secure supply of imports; and on the part of those who believed in the future of tropical forests, there was a confidence that the benefits brought by a flourishing timber trade would lead to the better conservation and management of those forests. Within ITTO, Objective 2000 became the practical expression of this belief.[9]

2

Sustainable forest management: a response to destruction

'To be strictly accurate, all that can be said about any management is that it is consistent with sustainability'

Introduction

Chapter 1 gave a brief survey of land use history and the economic and social forces that have shaped the distribution of forests in the past – forces that are now stronger and more pervasive than ever.[1] They are powered by many recent developments, among them the growth of human numbers, rising expectations, supra-national concerns such as biological diversity and climate change, and the many elements inherent in an increasingly globalized world. Modern societies require or desire many more things from forest lands than food, water and fuel. How can all these demands be accommodated? The standard answer is sustainable forest management – always an objective of ITTO, and since 1990 its Holy Grail.

The Tropical Forest Problem

Most people now believe that 'the tropical forest' is a resource of great value to humanity and that it is 'in danger'. This has become one of the clarion calls of environmentalists, a call that is now echoed in the statements of political leaders in both the developing and developed world. But there is still much misunderstanding of the nature and scale of the problem and, above all, of what might be done about it.

First, though, it is necessary to find out what the real issues are, to separate what is important from the verbal froth. How special is 'the tropical forest'? How serious is its loss or degradation? How vital is it to preserve forest cover in the tropics? How much forest should be preserved intact and where? Is it possible to use some parts of the forest for economic purposes while maintaining environmental values? How might this be done? Where have planning and management failed? What are the reasons for failure and success, and what can genuinely be done to solve these problems in the prevailing political and economic milieu of the tropical countries in which the forest is found? This book will try to cast some light, however tentatively, on some of these questions; in particular, it will examine the role of an international organization, ITTO, in improving the status and condition of tropical forests.

There is a great deal of loose talk about 'the tropical forest'. So many hectares (or 'football fields') of 'the tropical forest' are disappearing every day (hour or minute); 'the tropical forest ecosystem' is the richest in the world; 'the tropical forest' contains x per cent of the species on the globe; the removal of tropical forest' will cause widespread climatic disruption. These pronouncements conceal as much as they reveal, for they suggest that 'the tropical forest' is a monolithic entity. In fact, there are very many different types of tropical forest, perhaps as many different ecosystems as there are in all the other forests of the world, and very few of the statements made in regard to the forest specify to which type or types they refer. This is not to belittle the problem but to emphasize that it needs to be articulated with much more precision. It also underlines the difficulties of global generalizations and the dangers of using these without taking account of all the qualifications that surround them.

Variety

Tropical forests are very varied. True rain forests are found in the wetter equatorial and tropical regions. These are very rich in species, frequently occur on poor soils and, if cleared or mismanaged, can prove to be very fragile. Destruction or damage are being caused by agricultural expansion, planned or unplanned, by bad forest exploitation, and recently by fire. Cutting wood for fuel is rarely a problem until the forest has already been reduced to remnants. As the climate becomes drier, the rain forests give way to semi-deciduous forests, then to deciduous forests which gradually become smaller in stature as the rainfall decreases. These drier forests have been much more extensively modified by agriculture, grazing and burning than the rain forests. They have been replaced over large areas by savannas, grasslands, degraded scrub or wasteland. In such areas, a shortage of fuel wood is often the main problem. The mountain forests near the equator have characteristics that are not shared by those outside the tropics (strong daily fluctuations but small seasonal differences), but mountain forests in the sub-tropics have much in common with the forests of more temperate regions.

There have been many attempts to produce global classifications of vegetation (including the forests of the tropics). Some of these have been based on ecological differences (in climate, soils or both); others have used the structure or physiognomy of the forest. Each has value for particular purposes. The recent study entitled *State of the World's Forests 2001*, by the Food and Agriculture Organization (FAO) of the United Nations, for example, distinguishes 11 ecological zones in the tropics, against which it plots data on deforestation.[2] These are: tropical rain forest; tropical moist deciduous forest; tropical dry forest; tropical shrubland; tropical desert; tropical mountain system; sub-tropical humid forest; sub-tropical dry forest; sub-tropical steppe; sub-tropical desert; and sub-tropical montane.

Within each of these broad categories of forest, however, are many different types related to biogeographic history and to local differences in climate, geology, landform and soil.[3] It is these smaller distinctions that are important if nations are to make the best use of their resources. For example, some forest types are rich in valuable timber species; others have few or none. The same applies to biological diversity. To be valuable for these purposes it is necessary to have some knowledge of the actual species present in each forest type.[4]

Most generalizations break down in the face of this variety. Take deforestation as an example.[5] There has been much debate about its causes: proximate causes such as deliberate government land clearing, better accessibility after harvesting, fire, and agricultural encroachment; and underlying causes such as poverty, population pressure and debt. In fact, although some general patterns emerge, the causes are almost always site specific and can only be identified by detailed diagnosis on the spot.[6] The effects also are site specific. In the case of biological diversity, some types of forest are very special; they are exceptionally rich in species, many of them endemic; and they may occupy quite small areas. If such sites are deforested, the effect on biological diversity is very serious. But the loss of the same amount of other kinds of forest would have little or no effect on biological diversity, for there are many types of forest which occupy large areas and are relatively poor in species. It is for this reason that it is unwise to predict the number of species that will be lost if a certain percentage of forest is destroyed; besides, number is not necessarily the best indicator of the extent of the loss. But identifying the distribution of biodiversity is of the greatest importance when selecting areas for protection.

Land Use and Management – the Two Elements of Sustainability

Some history

The onslaught on the tropical moist forest began in the late 1950s and accelerated in the 1960s. At first, this was mainly transformation of forest into agricultural land, often in large, official land-settlement schemes. Some were carried out well, others badly.

Shifting cultivation was also a cause of forest degradation. It, equally, could be carried out well or badly – depending upon the numbers of people, the fertility of the site, the length of the cycle and the farming methods and crops used.

At that time, the problems of the moist tropical forest were largely seen as problems of *land use*, not of *forest management*. The loss or degradation of forest by timber exploitation was not yet seen as an issue.

These questions of land use were widely discussed in the early 1970s, when FAO published its first forest resource assessment and also played an important part in the deliberations of the First World Conference on the Human Environment in 1972 (the Stockholm Conference). Although much of the concern at that time was expressed by the developed countries, this was not (as has sometimes been represented) motivated by a selfish concern to protect temperate forest resources, but by a genuine desire to persuade tropical countries not to commit the same mistakes as had been made by many developed countries in the course of their development.

Immediately after the Stockholm Conference, there was much international activity. In 1974, the International Union for Conservation of Nature and Natural Resources (IUCN) held conferences in Caracas, Venezuela and Bandung, Indonesia, on 'guidelines for tropical forest management'. These resulted in the publication of *Ecological Guidelines for Development in Tropical Rain Forests*.[7] In November 1976, FAO devoted the fourth meeting of its Committee on Forest Development in the Tropics to 'the values of tropical moist forest ecosystems and the environmental consequences of their removal'.[8,9]

The emphasis then was still on deforestation and therefore on questions of land use. It is interesting to look back on the contents of the IUCN guidelines. There were sections dealing with: land use policy and the allocation of land to various uses; allocation, conflict and multiple use; preservation of natural ecosystems and genetic resources; protection forests; timber production from natural forests; indigenous communities and shifting agriculture; water resources; transformation of natural forest into field and plantation crops; management of fisheries in river systems; pest control; and settlements, engineering works and industry. This was a quarter of a century ago. The same problems are with us today.

It is notable that the section on timber production in these guidelines was very short; nor did either of these two documents recognize the timber industry as any threat to the ecological integrity of tropical rain forests. Up to this time, harvesting had been on a small scale and was usually well regulated, and harvesting damage was slight. And so it was still assumed that the management of forests for timber would be professional and conform to the silvicultural practices of sustained yield.[10]

It was only when logging accelerated in the latter half of the 1960s that forest management for timber began to have serious side effects. Quickly, exploitation increased much faster than the capacity of forest departments or forest managers to control it. Large areas of forest, especially in Southeast Asia, were stripped of all their valuable timber with little consideration for safeguarding future stocks. Nations and individuals became rich on the profits. Tropical forests now faced two threats: bad land use planning and bad forest management. The former had to be treated at the geographical level (national, sub-national and catchment), the latter at the level of the individual forest.[11]

What came to be known as the sustainable management of a nation's forests now had two components – the planning of land use (how to use each parcel of land for the purpose, or purposes, that provided the best lasting social advantage) and the management of its forests (how to manage each parcel of land both efficiently and sustainably for its chosen purpose).

The concept of sustainable management – then known as conservation – was first clearly articulated in the *World Conservation Strategy*.[12] Its Introduction dealt with 'living resource conservation and sustainable development' – the first use of the latter term. Conservation was defined as:

the management of human use of the biosphere so that it may yield the greatest sustainable benefit to present generations while maintaining its potential to meet the needs and aspirations of future generations. Thus conservation is positive, embracing preservation, maintenance, sustainable utilisation, restoration, and enhancement of the natural environment. Living resource conservation is specifically concerned with plants, animals and micro-organisms, and with those non-living elements of the environment on which they depend. Living resources have two important properties, the combination of which distinguishes them from non-living resources: they are renewable if conserved; and they are destructible if not.

The *World Conservation Strategy* discussed, in turn, the maintenance of essential ecological processes and life support systems; the preservation of genetic diversity; and the sustainable utilization of species and ecosystems – a balance comprising patterns of land use, sustainable use, intensified use, protection and restoration.

Land use planning and the balance of uses

The guidelines of the 1970s tried to provide some guidance on the contribution that land use planning could make to the sustainable management of tropical forest lands. Thus, forest lands should be allocated, as appropriate, to meet each of the following uses:

- *As living space for forest-dwelling peoples.* The importance is judged by the fact that local people actually live there (especially if this has been for a long time), the materials they can obtain from the forest and their 'sense of place'. For shifting cultivators, the crucial factors are soil fertility and recovery after cultivation; for hunter-gatherers, the presence of particular species.
- *For harvesting of economic products, timber and non-wood forest products.* The importance is measured by the presence of valuable species, the quantity of a first crop, the potential for later crops, ease of extraction, the presence of labour and access to markets.
- *For biological diversity.* The importance is measured by such features as the richness of species, the numbers that are endemic, and the areas' significance as feeding, breeding and staging posts for particular animals.
- *For catchment protection and water supply.* The importance depends upon the position of the forest in relation to the headwaters of streams and rivers, the nature of its vegetation cover and the steepness or erodibility of its soils.
- *For climate control.* Two elements seem to be of particular importance: the albedo (reflectivity) of the tree cover, which governs how much water is transpired by the forest, and the amount of carbon stored in an area of forest and its soil (its value for withdrawing carbon from atmospheric circulation).
- *For conversion into agricultural land.* The importance for agriculture, whether for local use or for export crops, lies in the suitability of the soil and climate for the chosen crops. Access to markets is also important.

Tropical forest conservation should be concerned, first and foremost, with finding an acceptable balance between these different uses and intensities of use. How much land should be used in each of these ways? Are there limits beyond which it is unacceptable to go? If the prime objective is to retain the *potential* of forest lands to produce varied goods and services – a basic principle of sustainable management – a cardinal rule should be to give first priority to catchment protection and the conservation of areas of particular importance for their biological diversity. The reason – damage to either of these values is impossible, difficult or very expensive to remedy.

These considerations were to form an important part of the ITTO guidelines, the criteria and indicators, and the priority afforded to setting up and giving security to a nation's permanent forest estate (PFE).

Forest management

What is sustainable management?

Once a forest is destined for production, how should it be managed sustainably? Sustainable management is a beguiling term and open to many interpretations. It contains many uncertainties and ambiguities. Let us look at some of these.

The sustainability of management of any area of forest can be assessed from two different points of view related either to the desired product or to the state of the forest: (1) to maintain the potential of the forest to provide a sustained yield of a product or products; and (2) to maintain the forest ecosystem in a certain desired condition (often linked to the services it provides). Consideration of these two will sometimes lead to the same result, but not always! Both, however, depend upon further definition and clarification: in the first case it is necessary to answer the question 'what product(s)?'; in the second we must ask 'what condition?' This point can perhaps best be illustrated by examples.

What product?

The most natural interpretation of this question from the point of view of the timber trade is the sustained production of specified timber products. It should be possible to produce a silvicultural system to meet a certain mix of products, but if the aim is to maintain the potential to meet a market that may have changed during the period of the rotation, the system may have to be modified to cope with this uncertainty. There is also an important distinction between harvesting the mean annual increment of a certain chosen species and maintaining the potential of the forest to provide a sustained yield. The situation becomes even more complicated if the yield of other forest products such as rattan or game is to be taken into account.

What condition?

Environmental arguments are likely to concentrate more on the condition of the forest than on its products. These arguments will hinge on the condition that it is desired to maintain, and therefore on the proper position of any particular forest in national policies for the allocation of forests to different uses. For example, at one extreme it may be considered sufficient to preserve the 'forest condition' – the function of the forest in relation to soil conservation and water circulation. In this respect, a well-planned rubber plantation with cover crops, or even certain kinds of grassland, may be an acceptable substitute for natural forest; any systems of extraction and silvicultural management that provide adequate soil protection and continuous vegetation cover will equally prove acceptable. At the other extreme, if the purpose of management is to maintain the forest in as near as possible to a virgin state for genetic conservation and scientific study, this condition is only sustainable by leaving the forest untouched.[13]

Between these two extremes are many intermediates. For example, if the successful establishment of a certain commercially valuable tree depends on birds that are themselves dependent on a tree with no commercial value, the forest can only be maintained by natural regeneration if the second species of tree is permitted to remain (it would also be maintained by the artificial planting of the first species). Different treatments, too, may be required for forests at different stages of an ecological succession or which depend on natural events such as fire or cyclones.

Thus, if one is to be strictly accurate, sustainability in a particular area of forest can only be defined in relation to a specified set of products and a specified condition. It may, however, be possible to design a system that is an acceptable compromise between a number of objectives. For example, a 'minimum intervention' system of silviculture may provide a reasonable and sustainable harvest of timber and retain a very considerable degree of the biological richness of the forest; but the more intensively a forest is managed for timber production the more its composition is likely to be simplified and impoverished in its variety of species and genotypes. (Even 'high-grading', a low-intensity but highly selective style of harvesting, is considered by many to cause the genetic deterioration of the stock.) One alternative course of action would be to compensate for the intensification of timber production in some areas by more generous provision for species conservation in others; another possibility would be low-intensity production throughout.

This ambiguity in the definition of 'sustainability' and the need for a nation to have forests which meet a number of different objectives emphasize once again the cardinal importance of careful land allocation in land use policy and the need for an exact specification of management objectives in relation to each tract of forest.

Indeed it is questionable whether one, universal definition of 'sustainable management' is useful, because it will lend itself to different interpretations by different interests. To avoid misunderstanding it is essential to be very clear which kind of 'sustainability' we are talking about in each instance.

There are also wider questions relating to sustainability which should not be ignored. It seems certain that this century will see some degree of global warming. How this will affect the tropical moist forest is not clear, but it is evident from the forest fires in Kalimantan, Sabah and Brazil that even small climatic fluctuations can make forests which have rarely, if ever, burnt before, much more vulnerable to fire.[14] This should be taken into account in determining standards of management.

Another issue of great significance will be the selection and management of protected areas against a background of climatic change. The near certainty that this will take place should clearly have a considerable effect on the criteria used for selection, the choice of boundaries, size, internal ecological variation, and the relation between the management of the protected area and that of the areas that surround it. For example, if changes in temperature are likely, the ideal protected area would include a range of altitudes – in order to allow the migration of species upwards or downwards – or even, if the protected area is sufficiently large, a range of latitudes. If there should be changes in rainfall, a range of soils from wetland to dryland would equally provide for migration.

We have seen that it is relatively easy to apply the concept of sustainable management to a precisely delimited parcel or tract of forest being managed for a single purpose – the sustained yield of timber from a forest reserve, for instance – although there may be complications even in this simple application of the term. But it is much more difficult to apply the concept to examples of multipurpose management; almost inevitably, some trade-off must be made between the various objectives, which leads to the question: how much trade-off is acceptable?

But, even supposing that one is successful in resolving the conflicts and trade-offs between different objectives for the management of a forest area, there still remains the problem that no tract of forest is an island unto itself; decisions relating to it are nearly always influenced by external factors – its sustainability is not self-contained. Consider the following examples:

- If a large area of virgin forest is deliberately left untouched for the sake of its biological diversity and other environmental values, its condition would normally be considered 'sustainable'. But is it still sustainable if this forest is subject to damaging atmospheric pollution from outside, or if it is surrounded by a poor and land-hungry population that might be tempted to encroach into it?

- A forest reserve may be managed by the best silvicultural techniques for a sustained yield of timber, yet changing markets may make it necessary to move towards over-exploitation or abandonment. Is its overall management to be considered sustainable?

- An intensively managed plantation with appropriate inputs of fertilizers and pesticides is normally considered a sustainable form of land use in itself, very similar to the cultivation of an agricultural crop, but its sustainability is subject to the price of energy, chemicals and labour, and to changing public opinion.

- An area of forest is cleared for a sustainable agricultural development that is part of a well-considered national land use plan and is clearly for the benefit of the local people who surround it. Should the timber originating from this forest clearance be deemed to be the product of sustainable management?

These are not imaginary or unrealistic examples. Many such tough problems face anyone who tries to come to grips with the practical application of the concept of sustainable forest management.

What is the sustainable production of timber?

When primary forest is first logged it normally contains a large standing volume of timber, a variable proportion of which is marketable depending on composition and market demand. Because this standing volume has accumulated over a long period, the commercial timber is likely to be of a quality and volume that probably will not be matched in future cuts (because it contains slow-growing specimens and species, large diameters, etc), unless the logged forest is closed to further exploitation for a century or more. In this sense the first crop is not repeatable.

If the production of timber is to be genuinely sustainable, the single most important condition to be met is that nothing should be done that will *irreversibly reduce the potential of the forest to produce marketable timber* – that is, there should be no irreversible loss of soil, soil fertility or genetic potential in the marketable species. It does not necessarily mean that no more timber should be removed in a period of years than is produced by new growth; overcutting in one cycle can, at least in theory, be compensated by undercutting in the next or by prolonging the cutting cycle.

But even this description of sustainable production begs several questions. For example: markets will certainly change between one phase of logging and the next – new species will become marketable and fashions may decline. So the 'timber production' to be sustained will not be the same from one cutting cycle to the next; it will contain a different mix of species. One consequence of this is that valuable species may initially be over-exploited for economic reasons.

Moreover, the forest will certainly alter somewhat in composition as a result of selective harvesting and there may even be some loss of soil fertility. But potentially damaging changes of this kind can be compensated for by new investment – in these instances, by the application of fertilizers or by enrichment planting. Whether to do so or not is essentially an economic decision.

Generally accepted silvicultural practice requires the calculation of the volume of timber which may be cut in one year in a given area – the 'annual allowable cut' (AAC) – a volume which should be set at a level that provides the maximum harvest while ensuring that no deterioration occurs in the prospects for future sustainable harvests. When an area of virgin forest is cut, the AAC depends on the volume of marketable timber in the area which may be cut while leaving enough stems on the ground for the next crop. This is calculated from an inventory of the standing stock and an estimation of the length of the cutting cycle – the period between successive loggings. If the forest is well stocked, this figure may be high. But the situation alters for subsequent cuts, because the rates of growth of the remaining trees change when the first crop is removed. At this later stage it should be possible, by measurement of permanent sample plots, to determine a stable AAC which corresponds to the annual growth in volume of the forest in question. This is likely to be different from, and often but not always lower than, the AAC from the primary forest.

The same considerations governing harvest and replacements should, of course, also apply to other forest products, both plant and animal.

There is sometimes also confusion between sustainable production per unit area and sustainable supply, because of the use in trade statistics of 'production' in quite a different sense from that defined above to mean, in effect, the supply of timber from whatever source. Sometimes 'sustainable timber production' is used to mean continuity of supply from the natural forest as a whole, implying that when one source is exhausted, another will be found. It need hardly be remarked that this usage is dangerous, for it need not include any provision for continuity of production on exploited sites and can lead to the total destruction of the resource. In fact, in this sense, supply is sustainable until it runs out; then it is someone else's problem.

Intensities of management

Management in the broadest context can be defined as taking a firm decision about the future of any area of forest, applying it, monitoring the application and, subsequently, fine-tuning based on the results of the monitoring.[15] This management may be for various purposes (for production of timber or non-wood forest products, for protection, etc); the important point is to be clear about the objective and to stick to it! It can also be at different levels of intensity (see Box 2.1).

It is important to recognize that different levels of management may exist, at least potentially, in any country, and that all may be justifiably termed sustainable. On the other hand, any system, however good, may produce results that are unsustainable if it is not applied conscientiously and consistently. Many good systems have failed because they are so complicated that there is neither the will nor the ability to operate them properly for even a few years, far less for the duration of a scientifically based cutting cycle or even the length of a rotation.

To be strictly accurate, all that can be said about any management is that it is *consistent with sustainability*. By the same token, practices not consistent with sustainability can often be made so by small modifications.

Wider Considerations

The above discussion of sustainable management is mainly concerned with ecological (or environmental) sustainability – maintaining the potential of the land or forest resource to yield benefits. This was the subject of the early guidelines and, certainly, ecological sustainability is fundamental to sustainable management. But the concept of sustainability has expanded to include a number of other parameters.[16] These are concerned with socio-economic and managerial features and are different in kind from the ecological parameters. While the ecological features are timeless, social preferences and economic and

BOX 2.1 LEVELS OF INTENSITY IN FOREST MANAGEMENT

Tropical forests can be managed for the sustainable production of timber (or of non-wood forest products and services) at different levels of intensity. This is often misunderstood and it is assumed that, if the forest is not being managed intensively, it is not being managed at all. We may take five different levels as examples, starting with the lowest:

- *Wait and see*. Where forest is remote and there is, as yet, no market for logs, the most effective management may be to demarcate the forest and protect it from encroachment until it becomes worthwhile to extract timber.
- *Log and leave*. After logging, the forest is closed and protected from encroachment or further logging. The speed of recovery and the volume of the future crop will depend on the kind of forest, the nature and standard of the first logging and the length of time that the forest remains closed.
- *Minimum intervention*. Marked trees (up to the limit defined by the AAC) are removed with minimum damage to the remaining stand and according to a well-researched silvicultural system, leaving behind an adequate stock from which the next crop can be taken after a defined number of years. Removals are confined to stems intended for sale and defective stems of marketable species; no other species are interfered with. The area is then closed and protected without any further tending until the next logging is due.
- *Stand treatment*. Logging is carried out in the same way as in the previous system, but the growth of the remaining stems or regeneration is enhanced by various treatments, which may include the poisoning of unwanted stems or species, poisoning of lianes, opening the canopy, weeding, etc;
- *Enrichment planting*. The treatment is the same as either of the two above, but saplings of desirable species are planted where the stock of residual stems is low; special treatment is given to encourage these saplings.

Any one of these options, if properly applied, would constitute management for sustainable timber production. Both the benefits, in terms of timber sales, and the costs, in terms of protection, tending, planting stock and chemicals, increase from the first alternative to the last. The decision about which to use at any one time is essentially a matter of policy, and that is likely to be strongly influenced by prevailing costs and potential benefits.

Source: adapted from Poore et al, 1989.

managerial conditions are constantly changing and altering the milieu in which the ecological system has to function.

Thus, if solutions are to be lasting, they must be economically and financially viable and socially acceptable. It is necessary to develop towards the ideal in steps that are practicable. The development of a policy and the adoption of the measures necessary for tropical forest conservation depend upon the evolution of appropriate social and economic mechanisms. At the national level, there is a need for measures to be politically acceptable and to sit readily in the framework of sensible economic policies. At the local level, patterns of land use need to be developed that are beneficial to the people they affect and do not intrude harshly into the harmonious development of local communities. They should be based on a broad understanding of the ecological circumstances and the social setting.

There are many obstacles to this harmonious development. Central, perhaps, is a lack of understanding that the intact forest and its soils are a capital resource. This is reflected in the plundering of this capital and a failure to invest in its protection and maintenance. But there are many others, including: political pressures; social unrest; considerations of national security; a failure to treat land use questions in a manner that is integrated and socially sensitive; a lack of trained staff; and a lack of knowledge derived from well-conceived research.

There are many different starting points and progress will inevitably be uneven.[17] Some fortunate countries have the resources to plan immediately for the ideal. Others may plan to depend on timber and on the cash crops that can be planted on good forest soils to power their economic development; they should be encouraged to examine how far these objectives are consistent with the ideal – and therefore how far they are sustainable – and to move towards an ideal pattern of land use. Where domestic pressures to meet everyday needs are urgent and immediate, it may be totally impracticable to think in terms of all elements of the ideal; in these cases soil and water conservation, and the growing of food, fodder and fuel, are bound to be preferred to genetic conservation. This is an area where international planning may bridge the gap; it may be possible to preserve the same resources somewhere else or, if there is no other alternative, to compensate for the national hardship caused. There are cases too where the sustainable use of resources is likely to remain a pipe dream unless there is international investment; some countries haven't even the capacity to undertake land use planning, let alone implement such plans.

The trade and the future sources of its timber

Tropical timber can come legally from a number of sources: (1) previously unlogged forest which it is intended should remain as forest; (2) logged forest which is being logged again, with or without a plan of management; (3) unlogged or logged forest which is being converted to another use in either a planned or unplanned manner; (4) secondary regrowth and enriched secondary forest; (5) trees planted in association with agriculture, roadside and canal-bank trees, etc; and (6) forest plantations. Both (5) and (6) have potential, but at present make a very small contribution to tropical timber supply.

There are many uncertainties about the future relative importance of these categories, and the boundaries between them are by no means distinct. For example: (a) much forest that is planned to remain as forest does not in fact remain so; (b) there is still too little logged forest under a plan of management that is effective and strictly applied; this forest is therefore likely to be deteriorating in quality and potential productivity. Re-logging is generally determined by the availability of a market rather than by considerations of sustainable silviculture: if light, it may be sustainable, but, if not, it is likely to lead to progressive degradation of the resource; (c) clear felling, carried out when the forest is being converted to another use, results from what may be a perfectly legitimate land use decision, but permanently removes the forest from timber production. Moreover, the timber felled during clearing often does not enter the market but instead is burned; and (d) there are no accurate trend statistics about what is happening to degraded forest or to sparsely populated or abandoned land. According to circumstances, it may degrade further or recover naturally and become capable of providing a supply of fast-growing, light-demanding timber species.

As the supply declines in any country from the logging of land which is designed for conversion to agriculture and from the first cut of previously unlogged forest, the future must lie in five possible sources: managed natural forest; managed secondary regrowth; agroforestry; plantations; and external sources (imports). Meanwhile, the immediate international market reaction to shortages or very steep price increases in timber from traditional sources is likely to be a movement away from the countries where supply has declined to those which have a largely untapped forest resource, or to plantation timber. The temptation for these newly favoured producing countries will be to follow the downhill path of resource mining that has been pursued by many of their predecessors, largely because they have a resource which they consider to be infinite or because they see the process as capitalizing an expendable natural asset.

As previously unlogged forest is logged and forest designed for agriculture is progressively cleared of timber, the question arises whether the supply of tropical timber can be sustained at expected rates by tropical plantations alone, however efficient.

Either the management of natural forest for the sustainable production of timber must be practised widely over an extensive forest estate; *or* unprecedented steps must be taken to promote production in secondary forests, in plantations and in farm forestry – or to import. The decision about what should be the right mix is one for each individual producer country; but the nature and sum of these decisions is of considerable moment for the timber trade and for each individual producer country.

The joker in the pack is the possible effect of the large amount of timber which is beginning to enter the market from maturing plantations in temperate and sub-tropical regions (eg from New Zealand, Australia, South Africa, Chile and the southern US) and from the opening up of natural forests in Russia and Canada. This will compete directly with tropical timber in many of its present end uses.

The national economic context of forest management

There seems to be general agreement that most governments seriously underestimate the *economic* worth of forests (both as productive resources and for the services that they provide) and that, on the other hand, they do not appreciate the cost of transforming the capital of natural forest into other forms of capital. There seems to be little doubt, at least in the view of one school of economists, that if the full non-market benefits of forests were to be taken into account, the 'sustainable utilization and conservation' of natural forests would prove convincingly economic.[18] In fact, many of the so-called 'serious' economic valuations show that forests actually have surprisingly little economic value for anything but timber and perhaps carbon.[19] Yet, despite many attempts to include these non-market benefits in national accounts, politicians in all countries continue to make decisions on the basis of their political judgement, which often runs contrary to the economic valuations argument. This underlines the need to make sustainable forest management politically attractive through good public relations and by providing the public with accurate information. It also shows that arguments based on theoretical economics are often less persuasive than those based on real-world financial considerations.

The very great differences in the way in which forests are treated in *financial* terms make it difficult to determine whether sustainable forestry regimes are financially sustainable – and financially competitive with other land uses. At one extreme – and this situation is relatively common in Asia – profits from the sale of timber are expected to generate large amounts of foreign exchange for investment in national infrastructure or other forms of profitable enterprise; this is in addition to covering the cost of the government forest service and providing funds devoted, in theory at least, to reforestation and forest management. In Indonesia, for example, timber was the largest earner of foreign exchange in the 1970s.

In other instances, the exploitation of tropical forest provides profit for all involved in logging and the wood industries, but the cost of the forest service is met from general revenue, justified, presumably, on the grounds of protecting the supply of a valuable product, the employment provided by the industry and the environmental benefits of retaining the forests. In between there are a number of variants: separation of functions between forest service and forest enterprise; different methods of partition of royalties, with huge differences in the amounts charged; different ways of providing (or not providing) for the costs of forest management, etc.

Under these varying circumstances it is nearly impossible, from the information available, to judge whether natural forest management is financially viable (or indeed to assess what is meant by this term!). Yet, unless in the long run it is more profitable than other alternative uses of the land, it is unlikely to be widely practised.

The social context

The regulation of forest management has, in many countries, been considered to be exclusively the province of government, with exploitation being carried out by government itself or by contractors licensed by government. In carrying out these functions the customary rights of local peoples have been respected, or not, to very different degrees.

It is generally true, however, that the benefits from government-promoted forest exploitation have seldom accrued to those who live in or near the forest. For hunter-gatherers, timber extraction may seriously damage their environment, though the light extraction in some African countries apparently may have the opposite effect. But it is often the case that farming peoples seldom see the advantages of retaining forest as forest rather than converting it to another – to them more lucrative – use. The sustainable management of government-run forests depends, therefore, upon the resolve of governments, not only to manage the forest consistently for long periods, but to ensure that the people who live in or near the forest benefit from (and are preferably involved in) the management of that forest.

But there are other possible models for sustainable management. Small-scale, local successes in forests managed by private commercial concerns and by local communities suggest models that might serve much more widely. What is crystal clear is that no system works unless there is long-term security for the managed forest and unless those concerned can see that the enterprise is profitable to them. It must also, of course, be profitable to the country.[20]

Conservation – the moving target

If the resources of tropical forest lands are to be effectively and lastingly conserved, it will not be enough to design answers that only fit the circumstances of today. There will be great changes even in the next few decades; and certainly by the end of this century. While conservation is concerned with attaining stable balances, development is dedicated to change. Policies should be imaginative and designed to adapt to changing circumstances: in the balance between populations and resources; in economic well-being; in relation to climate change; in world energy policies; in the balance of trade; and in the attitude of people to environmental issues. Many land use policies are obsolete before they are implemented. Conservation policies for tropical forest lands must, therefore, look forward, be integrated as far as possible with policies for population and for all sectors of the economy, and aim to hit a moving target.

International dovetailing

There is an important international element in policies for tropical forest conservation. This has particularly gained prominence since the landmark United Nations Conference on Environment and Development (UNCED) was held in 1992. The effective conservation of tropical genetic resources is a matter for international planning and cooperation in measures for the choice and safeguarding of protected areas. The same is true of measures to protect species. But this in no way lessens the duty of each nation to manage its own resources in a responsible way – indeed it enhances it by adding an international dimension.

But good land use planning and proper sustainable use can become very difficult unless international policies encourage these, or at least do not positively discourage them. International markets and policies can very readily encourage unwise use of resources: the movement into cash crops rather than food crops; incentives to over-exploit tropical forests to satisfy foreign markets; and the instability of commodity markets leading to unwise decisions about land use. Moreover, although difficult, it is important that international policies for tropical forest conservation, particularly of trade and aid, should provide a real incentive to the sustainable use and conservation of the resource.

As the world becomes increasingly globalized, the need for international harmonization has become more urgent. International agreements have proliferated, national trade and aid policies have changed and the roles of international organizations have evolved, even that of the United Nations (UN) itself. Among the intergovernmental organizations, only ITTO focuses on the sustainable management of tropical forests and a sustainable tropical timber trade. The next chapters consider how and why ITTO came into being, how it has operated, what its achievements have been and how important a player it has been – or could be.

3

Genesis of a treaty: the ITTA takes shape

'To encourage the development of national policies aimed at sustainable utilization and conservation of tropical forests and their genetic resources, and at maintaining the ecological balance in the regions concerned'
International Tropical Timber Agreement (ITTA), 1983

Phase 1

The International Tropical Timber Agreement (ITTA)[1] was adopted in 1983 and came into force on 1 April 1985; but the idea originated many years before, in 1966, within the United Nations Conference on Trade and Development (UNCTAD). Established by the UN General Assembly in 1964, UNCTAD's stated aim is to help in restructuring the traditional patterns of international trade in order to enable developing countries to play their part in world commerce. At the time of UNCTAD's creation, increased trade, both with the industrialized countries and among developing countries themselves, was regarded as the key to the accelerated development of these countries and to a rising standard of living for their peoples. In 1985, UNCTAD had 168 member countries.

In November 1966, an UNCTAD/UN Food and Agriculture Organization (FAO) Working Party on Forest and Timber Products proposed, at its first meeting, that a tropical timber bureau should be established. The idea was revived at its second meeting in September 1968 with a recommendation that the bureau should indeed be set up. Accordingly, a proposal was prepared by the Secretary General of UNCTAD and the Director General of FAO. The focus at this time was totally on trade. The tasks of the bureau were to be the collection and exchange of information on markets and their requirements, and on the properties, uses and availability of tropical forest products.

At this stage, the United Nations Development Programme (UNDP) was approached, not least for the financing that it might bring in. As a result, it commissioned a special fact-finding mission to ascertain the views of producing (exporting) countries on the creation of a tropical timber bureau. Its report was considered by UNCTAD, UNDP, FAO and the International Trade Centre (ITC).[2]

Table 3.1 *Origins of the ITTA, 1983*

Phase 1 – International Trade Centre
International Tropical Timber Bureau

November 1966	First session of UNCTAD/FAO Working Party on Forest and Timber Products in Geneva. Establishment of tropical timber bureau proposed.
September 1968	Second session of UNCTAD/FAO Working Party. Recommended that a bureau be set up to concern itself with the collection and exchange of information on markets and their requirements, and with the properties, uses and availability of tropical forest products. Secretary General of UNCTAD and Director General of FAO submitted proposal to UNDP.
April–May 1969	UNDP-financed mission to West Africa to obtain views of governments and trade.
1969–73	Studies and consultations. FAO wanted to expand the originally conceived role to give greater emphasis to 'upstream' activities. UNDP withdrew its support; UNCTAD concerned at change of emphasis.
May 1974	Committee on Forest Development met in Rome, where tropical countries reaffirmed their interest in the concept.
July 1974	ITC reactivated the idea.
1974–6	CIDA-financed study to consult with tropical timber exporting countries about possible establishment of an International Tropical Timber Bureau.
September–October 1976	Intergovernmental Consultative Meeting of Tropical Timber Producing Countries giving general support to International Tropical Timber Bureau. Consideration of draft agreement and request to ITC to convene a Negotiating Conference.
October–November 1977	Intergovernmental Negotiating Meeting of Tropical Timber Producing Countries. Agreed text forwarded to Secretary General of UN.
16 January 1978	Agreement opened for signature at the UN HQ in New York (but never ratified).

Phase 2 – UNCTAD
International Tropical Timber Agreement

23–25 May 1977	First Preparatory Meeting on Tropical Timber under UNCTAD Integrated Programme for Commodities.
October 1977	Intergovernmental Expert Group Meeting on Tropical Timber
24–28 October 1977	Second Preparatory Meeting
November 1977	Tropical countries met to discuss agreement.
23–27 January 1978	Third Preparatory Meeting
31 July–4 August 1978	Fourth Preparatory Meeting
22–26 October 1979	Fifth Preparatory Meeting, Part 1
7–18 July 1980	Fifth Preparatory Meeting, Part 2
1–11 June 1982	Sixth Preparatory Meeting of 50 nations met in Geneva. Elements of the agreement finalized.
1983	Negotiating Conference on basis of Japanese draft, based on jute and rubber agreements.
1983	International Tropical Timber Agreement adopted.
October 1984	First possible date for the ITTA to enter into force.

Phase 3
International Tropical Timber Organization

March 1985	International Institute for Environment and Development (IIED) meeting
April 1985	The ITTA enters into force
June 1985	1st Session of the International Tropical Timber Council (ITTC)

The proposal simmered for the next eight years, during which time there were many studies and consultations, and divergent views were canvassed about the range of activities that should be covered by the bureau. FAO, for example, wanted to expand its role, as originally conceived, to give greater emphasis to 'upstream' activities (presumably forest management and timber extraction). Others considered that the proposed bureau should not concern itself exclusively with the timber trade, for 'it would be ludicrous' if the demand generated by the trade were to 'lead to indiscriminate felling'. Nevertheless, people were concerned that a greatly expanded role might create a new bureaucracy. The proposed expansion of function was seen as likely to build up a bureau 'whose function would be, from the beginning, highly ambitious and complex, require very considerable financing, and would risk duplicating the work of FAO, UNCTAD and UNIDO (United Nations Industrial Development Organization), as well as requiring a secretariat with extensive in-house facilities'.[3,4]

The proposal, by this time, was faltering. UNDP withdrew its support on the grounds that 'the concept of the Bureau [was] less clear now than before and the involvement of UNCTAD, as now proposed, [was] tenuous'. UNCTAD, for its part, was concerned that the original idea had been greatly altered, and that 'the more specific needs for expanding and rationalizing international trade in this commodity [tended] to be relatively less in focus that originally foreseen'.

By mid-1974 it had become necessary to restate the trade case for a bureau. Meanwhile, the fears of the timber trade had been allayed by making it clear that the proposed bureau would *not* become involved in commercial activities. This qualification was important in the eyes of the trade, which had feared that the tropical timber sector might be subjected to the theories and ideologies central to commodity agreements, with attempts to fix prices and to regulate the flow of products into markets. It was their view that the activities of the tropical timber trade were too complex to be treated in this way. Most important, however, the trade liked to have the freedom to react quickly to market conditions and to have the competitiveness that stems from the quality and performance of individual companies.

In July 1974 – eight years after the original proposal – the ITC reactivated the idea. A study financed by the Canadian International Development Agency (CIDA) was carried out to examine the possibility of establishing an international tropical timber bureau, the mandate being that the bureau was to be for producing/exporting countries only. In 1975, on behalf of ITC, the two persons entrusted with this study visited and consulted with the governments of tropical timber exporting countries in all three producer regions, as well as trade and research organizations in importing countries, in order to sound out their views. They reported that views were positive but, at the same time, indicated that it would be advisable that importing countries should also be members of the bureau.

At this time, the proposed functions of the bureau were to be:[5]

• to stimulate demand for, and further the use of, tropical timber, including the lesser-known species and the products manufactured therefrom;

- to collect, collate and disseminate technical information on various tropical timbers warranting promotion and carry out appropriate market development programmes in light of the requirements of the import markets and the existing facilities for these types of work;
- to develop channels for, and maintain, the free exchange of market intelligence and technical knowledge between importing and producing countries.

There was nothing about the management of the resource – yet. But some quarters of the trade were already interested in extending these objectives to cover the need for sustainable supply. This is illustrated by a quotation from a letter from Mr Geoffrey Pleydell on behalf of the United Africa Company to Om P Mathur of ITC:

> tropical forests have a key role to play in meeting needs for wood products. They are already recognised by producer nations as important potentially recurring natural assets capable of providing employment, revenue, valuable exports and contributions to domestic development. FAO statistics indicate, in the long term, an ever increasing demand for wood. It has been said that natural tropical forest plus properly managed regeneration, will help meet these demands... There are problems of individual producing nations and their need to maximise, on a sustained basis, the value of their own forests.[6]

In February 1975, ITC distributed an aide-mémoire about a project to establish an International Tropical Timber Bureau. This put the emphasis on commercial prospects arising out of research and development into lesser-known species, and on information about the need to reinforce trends towards further processing. It noted a lack of effective promotional activities, and weak communication between producers and consumers. The aide-mémoire saw the proposed bureau as helping to stimulate demand and use and to broaden the range of end uses. There would be a free exchange of market intelligence which would help to identify needs and inspire activities within the scope of the bureau to collect, collate and disseminate information and to investigate market opportunities and the standardization of grading.

As a result of the positive outcome of these consultations, ITC held two meetings. The first was the Intergovernmental Consultative Meeting of Tropical Timber Producing Countries held between 27 September and 1 October 1976. At this meeting, the producing countries gave general support to the creation of an International Tropical Timber Bureau. They considered that this bureau would be a useful instrument for sustaining the efforts by national and regional institutions to develop the market for tropical timber and timber products and, further, that a world-wide body representing tropical timber producers would be able to provide certain services more economically and efficiently than would national or regional bodies. This would be especially so in the collection and dissemination of market and technical information and in matters relating to lesser-known hardwood species.

This first meeting considered a draft agreement to establish an international tropical timber bureau and requested the ITC to circulate this draft to governments of tropical timber producing and exporting countries for their comments.[7] It also requested the ITC, after taking these comments into account, to convene a negotiating conference for the further consideration and adoption of the text of an agreement to establish an International Tropical Timber Bureau. At that time, it was thought that the new bureau would consist of three elements: an assembly; a technical committee; and a secretariat. Endeavours to bring in politically sensitive issues such as price control and buffer stocks were ruled out.

A second meeting – the International Negotiating Meeting of Tropical Timber Producing Countries – was held about a year later to adopt the text of an agreement. The negotiations proved difficult, partly because of the commercial rivalry between the African and Asian producing/exporting countries. Nevertheless, a definitive text was agreed and forwarded to the Secretary General of the UN to be opened for ratification, approval or acceptance at UN Headquarters by those countries that qualified for membership – to come into force when half of the countries invited to the meeting (ie 16) acceded to it. It seems that no country did sign this agreement. The reason might have been that UNCTAD's First Preparatory Meeting on Tropical Timber had already taken place in May. In addition, the fact that full membership was restricted to tropical wood exporting nations, with importing nations solely as associate members, made the agreement less attractive to the importing nations.

With the failure of this negotiation, Phase 1 of the evolution of ITTO was over. The idea of an international instrument for tropical timber seems to have ceased to be of policy or strategic interest to the ITC. But the interest of the producing countries in the creation of an international institutional instrument for tropical timber remained strong. This interest provided the impetus to the UNCTAD Preparatory Meetings that followed.[8]

Phase 2

Since 1976, UNCTAD's Integrated Programme for Commodities (IPC) had aimed to improve conditions in world markets for 18 primary products of particular export interest to developing countries; these included coffee, jute, rubber, sugar, olive oil, tea and tin. This Programme had many objectives, ranging from improved production of the commodity, through price stabilization, to the increased participation of developing countries in the processing, marketing and distribution of their commodities. A central feature was the creation of the Common Fund for Commodities to finance buffer stocks – an appropriate measure for most of the commodities included in the programme.

At the UNCTAD IV Conference, one of the African members proposed that tropical timber should be added to the list of IPC commodities – the last of the 18. Timber, however, was rather different from the other 17 commodities. The international tropical timber importers scorned the idea of applying to tropical timber the same recipe as for other commodities – the stabilization of prices by the purchasing, stocking and releasing of timber stocks by an international body – a view shared by FAO. Traders considered such stabilization to be totally impractical, perhaps impossible, because of the number of commercial species involved, the different categories of products, different quality specifications, different grading rules and the different origins of tropical timber. Furthermore, timber, especially logs, could deteriorate and lose value if held in buffer stocks for a long time.

Nevertheless, there was remarkable enthusiasm. To quote Ulrich Cording, the Economic Affairs Officer in UNCTAD (Cording, 1985):

> an unusual degree of goodwill and desire to reach agreement among governments from both producing and consuming countries emerged. This un-ideological negotiating climate and desire for commonly shared progress cannot be explained by the high commercial importance of tropical timber alone. It clearly reflects the recognition that tropical timber is a commodity unlike all others: harvested from mostly virgin forests as a product of highly fragile ecosystems, renewable under certain conditions only after a long timespan, and of paramount importance to mankind in so many respects – productive, protective and social.

The bulk of the international trade in tropical timber then was between the international importers in Japan (then the largest importer in the world in volume terms – mainly of logs), Europe (principally Germany, France, the UK and Italy) and the USA, and exporters in the three tropical regions – the Asian (led by Malaysia), the African (led by Côte d'Ivoire) and the Latin American (led by Brazil). The three producer regions had rather different priorities, which led to some commercial rivalry reflected in the UNCTAD negotiations. Malaysia and Indonesia were worried about over-logging, whereas the African countries wanted to improve the use of unlogged forests. Latin America, with an even smaller share of the international trade, initially sat on the fence, waiting to see whether accord would develop between Africa and Asia.

A new process began, therefore, within UNCTAD, with a fresh momentum. There was clearly some continuity of thinking between this new initiative and the ITC discussions and negotiations, but there is no official reference to the ITC initiative in the UNCTAD meetings. Six preparatory meetings, starting in 1977, led ultimately to a negotiating conference, the United Nations Conference on Tropical Timber 1983, at which the ITTA was finalized and adopted.

At the first preparatory meeting (May, 1977) the focus of UNCTAD was on establishing the scope and scale of the problems facing tropical timber,[9] and it was only at the second meeting (October, 1977) that the consumers, for the first time, voiced their concern for the continuity of supply – a reference made possible because the UNCTAD meetings, unlike those of the ITC, were not confined to producing/exporting countries. This second meeting focused on research and development and on marketing, but at the third (January, 1978) the question of 'upstream' aspects was raised by the consuming countries, discussed and remitted to the fourth meeting.

There was a sharpening of focus as the discussion continued. By the fourth meeting, the main topics had been narrowed to four: improved market intelligence; increased processing in producer countries; research and development; and reforestation and forest management. At the same time, the centre of gravity shifted. Environmental issues, arising from the alarming rate of deforestation, the inadequate rate of reforestation and the differing status of forest management in producing countries, only began to be discussed seriously in late 1977 and early 1978.

This environmental trend was evident at the fifth meeting (October 1979 and July 1980). Discussions were based on the four main topics, but most of the documentation was in fact concerned with reforestation and forest management.[10] Two groups of experts were convened to concentrate, respectively, on market intelligence and on research and development for tropical timber; and UNCTAD, FAO and the World Bank were required to submit a paper to the next meeting on the amount of reforestation required for industrial tropical timber and the costs and conditions to cover this.

Consumer countries insisted from the early stages of these preparatory meetings that it was essential, in the interest of international trade in tropical timber, that there should be continuity of tropical timber supplies. This was not surprising, because the delegations of the major importing consumer countries included experts drawn from associations of tropical timber agents and importers. The producers' attention, on the other hand, was more focused on markets: market access and the changing market structures in the main importing countries; the need for timely and pertinent market information; customs levies on processed tropical timber; and freight rates and allied questions.

The focus of the producers was understandable. Upward and downward swings of demand cause problems for both producers and consumers. High demand pushes prices up and more orders mean more felling for logs, lumber and plywood. It can take weeks to fell, extract, process and dry lumber. If, in the meantime, strong demand and high prices are overtaken by weakness in the market, producers can find themselves overstocked through lack of contracts and can see prices fall, sometimes to levels that have the potential to push individual companies out of business.

Price stability and a better balance between supply and demand are sought but are difficult to achieve in practice. Markets are always changing because the general economies of importing countries fluctuate and there is a huge range of competition to supply. Importers may fare slightly better than producers because they are more sensitive to change in their own individual markets and can react more quickly. Even so, they can find themselves overcommitted to outstanding supply contracts, overstocked and holding high-cost stocks in a low-price market. In times of high demand, importers may be tempted to over-order and exporters to take on more orders than they can realistically produce on time. In difficult times, importers may try to cancel existing contracts and exporters become low-price sellers. Successful assessment of market trends is a key element in the success of a timber company.

Undoubtedly, it was the desire to find ways to iron out unsettling imbalances in the production and consumption of timber which eventually led to the establishment of ITTO. And, of course, it rapidly became apparent that within tropical forested countries the balance between the capacity of industrial production and the sustainable capacity of the forest was fundamental to maintaining long-term benefits from both. Certainly, during the six years of preparatory meetings, both forestry (the growth of trees for production) and the ecological values of forests became more prominent. One reason for this was that the delegates to the preparatory meetings, particularly of producer countries, were mainly qualified foresters.

At some time during the course of the preparatory meetings, the Commodities Division of UNCTAD organized a workshop in Abidjan specifically for the producing countries, in order to exchange views and test attitudes. UNCTAD's Terence H'pay prepared a draft agreement for discussion at this workshop which included a proposal that there should be an international tropical reforestation fund to meet the needs of reforestation and forest management. This remained on the agenda of the preparatory meetings to be further developed. Some solidarity developed between the producer regions based on common interest and a proposal about the allocation of votes helped to allay fears by providing an initial allocation of 300 votes to each region which was not dependent on exports or forest resources.

The insistence by the consumers on the future of commercial supplies raised the question of inventories. These were provided by the Forestry Department of FAO, which, by then, had made great progress in assessing tropical forests. The results of this assessment led inevitably to discussions about the rate of loss of natural tropical forests and the causes of this loss. The subject of deforestation, including its possible effect on atmospheric carbon dioxide, aroused the interest of environmental groups. By this time, the preparatory meetings had clearly decided that they must give due consideration to ecological aspects of the management of tropical forests for timber. They still had reservations, however, that these matters properly lay outside the remit of UNCTAD.

The central themes in the preparatory meetings, therefore, moved from a concentration on tropical timber supplies and market intelligence towards including consideration of utilization, reforestation and conservation. New questions came to the fore: reforestation, its present deficiencies, its technical difficulties (such as that faced in reforestation with dipterocarps), the need for further research and the pressing requirement for more finance.[11] Discussions now ranged comprehensively over the whole question of tropical forest management, including reforestation. Observers from the World Wide Fund For Nature (WWF) and the United Nations Environment Programme (UNEP) raised awareness and helped to tone down the objections raised by many participants that environmental matters lay outside UNCTAD's competence and should only be discussed in those parts of the UN that had responsibility for them. At this stage, too, the chairmanship of the preparatory meetings was assumed by Dr Tatsuro Kunugi of Japan, who later became Chairman of the Negotiating Conference. His appointment was important, for Dr Kunugi was very alert to environmental issues and was the initiator of the environmental clause in the third paragraph of the Preamble of the ITTA:

> To encourage the development of national policies aimed at sustainable utilization and conservation of tropical forests and their genetic resources, and at maintaining the ecological balance in the regions concerned.

UNCTAD, for its part, had held a meeting of experts about the way in which any available funds might be used to fund a series of 42 projects in tropical Africa, Asia and Latin America – projects designed to promote more rational and economical use of wood, prevent deforestation and improve methods of logging, harvesting and training. Priority was given to projects on wood usage and schemes to protect and expand natural forests. It is notable that, even at this early stage, detailed consideration had already been given to priorities for funding. However, no disbursement could take place until the agreement was signed.

At the fifth preparatory meeting the proposal for a reforestation fund was unfortunately dropped after an intervention by the representative of the International Bank for Reconstruction and Development (IBRD) (an institution of the World Bank), who argued that there was no need for such a fund as the Bank could and would provide the necessary funds for regeneration and management.[12] Many of the later weaknesses of the Agreement stem from this failure to make provision for a secure source of funds; without independent untied funding – a necessary condition for action in the tropics – very little of consequence could happen.

The ground was finally ready for the negotiating conference at the sixth preparatory meeting in June 1982. Japan had tabled a draft agreement based on those for rubber and jute. The conference was organized by UNCTAD in Geneva and attended by 50 nations. Japan and the principal Southeast Asian producers played the leading roles. At this meeting, the four elements which would constitute an international tropical timber agreement were finally agreed. An UNCTAD/FAO paper on reforestation and forest management of tropical timber within the IPC was considered and the 42 projects that it recommended were organized into five programmes. It was decided, however, that the final choice of projects should be left to a responsible producer/consumer body within the framework of the agreement.

It must be recognized that many in industry and the trade were barely aware of the moves to set up an intergovernmental body or of its possible impacts and, if they were aware, probably saw it as a rarefied and remote high-level organization; they were more concerned with everyday supply and demand. But those in the international tropical timber trade who knew about these developments were comfortable with the proposed role of the new organization. Prospects of direct interference had faded and the direction was positive and supportive of industrial and market evolution. Some timber companies, who had a long and profitable involvement in log extraction and log exports, may have felt that moves away from that trade would make life more difficult for them, but they also understood that there would be time to adjust and plan for change. Some log exporters questioned whether sawmilling for export could be as profitable as shipping logs. 'Round is Sound' was a phrase used in West Africa to encompass an activity which minimized costs per usable cubic metre of output. By and large, the industry and the trade understood the pressures for change; and timber companies in the tropics could hope that the new organization would give them fresh opportunities to develop new business, especially as one of its aims was to encourage more processing at source.

The bigger picture, in some tropical countries, was an increasing awareness that forest resources were being eroded and that policies were needed to stabilize the forest and ensure the sustainability of its functions. This, however, must be accompanied by the development of an industry based on smaller harvests and more refined and efficient processing.

The stage was now set for the Negotiating Conference, which met in March and November 1983. Representatives of 70 countries participated (36 producers and 34 consumers), covering between them 93 per cent of all productive closed broad-leaved tropical forests and 98 per cent of the trade in tropical timber products such as logs, sawnwood, veneer sheets and plywood. Thus, finally, the ITTA, 1983 was adopted in the record short time of five weeks of negotiation.

The tropical timber countries broke new ground with the Agreement. In other commodity agreements, the votes in the Council were distributed in accordance with the share of world trade held by member countries. For tropical timber, a compromise had to be found between the existing exporting strength of the Asia-Pacific region and the potential future strength of the Latin American region, which would depend on the large area of its forests. The outcome was a completely new way of distributing votes. Of the 1000 votes held in common by the producing members, 400 were to be equally distributed between the three producing regions, with each region distributing the votes allocated to it equally among its members; 300 votes were to be distributed among the producing members in accordance with their respective shares of the total forest resources of all the producing members, and the remaining 300 in proportion to the average of the values of their net tropical timber exports during the most recent three years. According to Ulrich Cording, a considerable effort was required from consuming countries to accept this unusual solution in order to avoid deadlock in the negotiations (see Chapter 4).

The definition of 'producing member' was equally new. This was based on the possession of tropical forest resources, so that countries that did not export tropical timber were nevertheless recognized as 'producing members'. Thus, all countries might seek help for the maintenance and management of their forest resources and could draw on the combined knowledge and experience of all member countries. This was clearly an important move towards global cooperation and solidarity.

Unfortunately, as already mentioned, the proposal for a reforestation fund was not taken up, and this has proved a weakness of the Agreement ever since. Instead, projects were to be financed from a Special Account.

Phase 3

After the adoption of the Agreement in 1983, there was a delay in implementing its requirements on the part of both consuming and producing countries. The earliest date for definitive entry into force (1 October 1984) was not met; and, by early 1985, there was concern that even the deadline for 'provisional entry into force', 31 March 1985, would also pass.

By mid-February 1985, 15 consuming countries had signed the Agreement, of whom 13 had deposited an instrument of ratification. The target for consumers (663 votes) would be passed if the two remaining consumers, Spain and Italy, acted promptly. But the situation in relation to the producing countries was alarming. Ten producing countries carrying 500 votes were necessary. Six had signed (in order of signing, Liberia, Indonesia, Gabon, Honduras, Bolivia and Malaysia), but only Indonesia and Malaysia had ratified. Eight more instruments representing at least 235 votes were still required. According to Ulrich Cording: 'If this deadline is not met, there is a certain risk that the ITTA might never be established due to inaction on the side of the producing countries, since several consuming countries have indicated that they do not intend to set up an ITTA with only minimal producer participation'.[13] The future of the Agreement was on a knife-edge.

At this stage the International Institute for Environment and Development (IIED) entered the scene. During 1984, IIED's Brian Johnston had become interested in developing a programme directed towards the sustainable management of tropical forests. I joined him and took over when he left IIED later in 1985. We envisaged two elements in this IIED programme: to promote in-country studies in tropical countries to determine the effect of all government policies on the management of their forests; and to examine the influence of international trade and aid. We were especially interested in the potential of the ITTA, with its unusual environmental clause, to act as a powerful agent in promoting sustainable management. Accordingly, we determined to call an international seminar to stimulate the necessary ratifications before 31 March 1985.

The meeting took place in London between 8 and 10 March.[14] It was attended by 13 representatives of consumer countries and 15 from producers, three from environmental organizations (apart from IIED itself), four from the trade and four from FAO, UNCTAD and UNEP. Japan was especially well represented because of its great interest in attracting the headquarters of the new organization to Yokohama. Notable absentees were the US, the UK Timber Trade and the World Bank.

The seminar was addressed by representatives of Cameroon, Indonesia, Malaysia, FAO, UNCTAD, IIED, the Association Technique International des Bois Tropicaux (ATIBT) and the Dutch Timber Trade Association. After long and valuable discussions, the meeting approved a Statement to be sent to all governments which had been party to negotiating the ITTA.

The final paragraphs of the Statement read as follows:

This seminar, convened to explore the implications of the Agreement for the relationship between the utilization and conservation of the tropical forest resource, confirmed that these need not be contradictory goals. The participants of this seminar recognized that the major causes of degradation are the combined effects of population pressure, the extension of agriculture and the demand for fuelwood, rather than the utilization of forests for timber products. The benefits derived from the trade, however, could greatly contribute to the management and conservation of the resource which in turn ensure the sustainability of economic and social development.

Nevertheless, despite this consensus as to the potential value of the agreement, the seminar noted with grave concern that the requirements for entry into force have still not been met as the 31 March deadline approaches.

The seminar urges everyone concerned to do all in their power to see that the Agreement enters into force... The seminar stresses that the ITTA can rightly be regarded by the whole international community as a major success for international cooperation and solidarity.

Cables were sent. The necessary instruments were deposited in New York by 31 March, and the ITTA came into effect on 1 April 1985.

4

ITTO's early days: optimism and experiment

'It is new! It is global! There is no traditional or established way to follow. No scientific formulae to be guided by. No red tape to be strangled by... There is nothing to dismantle, nothing to change before it can start'
Dr Freezailah, Executive Director

The International Tropical Timber Agreement

The ITTA came into force on 1 April 1985. Its stated objectives were:

(a) To provide an effective framework for co-operation and consultation between tropical timber producing and consuming members with regard to all relevant aspects of the tropical timber economy.

(b) To promote the expansion and diversification of international trade in tropical timber and the improvement of structural conditions in the tropical timber market, by taking into account, on the one hand, a long-term increase in consumption and continuity of supplies, and, on the other, prices which are remunerative to producers and equitable for consumers, and the improvement of market access.

(c) To promote and support research and development with a view to improving forest management and wood utilization.

(d) To improve market intelligence with a view to ensuring greater transparency in the international tropical timber market.

(e) To encourage increased and further processing of tropical timber in producing member countries with a view to promoting their industrialization and thereby increasing their export earnings.

(f) To encourage members to support and develop industrial tropical timber reforestation and forest management activities.

(g) To improve marketing and distribution of tropical timber exports of producing members.

> (h) To encourage the development of national policies aimed at sustainable utilization and conservation of tropical forests and their genetic resources, and at maintaining the ecological balance in the regions concerned.

From the point of view of sustainable forest management, the most important elements contained in these objectives were the 'expansion and diversification of international trade in tropical timber', the assumption that the aim should be 'a long-term increase in consumption' based on 'continuity of supplies' and, of course, the all-important Clause (h).

Underlying these objectives was the unspoken conviction that the accessible parts of tropical forest would only continue to exist if they were used for an economic purpose – the most important being the production of wood; protection alone would not suffice. In fact, the whole edifice of ITTO was based on the hypothesis that it was technically and politically possible to manage tropical forest for a sustained yield of timber while maintaining its other values. One purpose of this book is to examine whether this hypothesis has or has not been borne out by subsequent events.

Under the Agreement, the Organization's 'highest authority', the International Tropical Timber Council (ITTC), was to consist of all the members of the Organization organized into two categories: producing and consuming. Each of these groups was allocated 1000 votes. Although Article 12 urged that 'the Council should take all decisions and make all recommendations by consensus' (and in fact this was an almost invariable rule in subsequent business), the dichotomy – normal practice in commodity agreements – led, as we shall see, to unfortunately confrontational behaviour.

The Organization was to act through: providing an effective framework; promoting, supporting, improving or encouraging various activities; and encouraging the development of national policies. It was to do this by discussion in Council sessions; the provision of services such as market information; and the approval, funding and management of projects. The present structure is shown in Figure 4.1; the early structure was very similar.

Three permanent committees were specified in the Agreement – the Committee on Economic Information and Market Intelligence, the Committee on Reforestation and Forest Management, and the Committee on Forest Industry. Their functions are fundamental to the operational role of the Organization. Those for the Committee on Reforestation and Forest Management are given in Box 4.1

As it turned out, projects were to play a very prominent part in the deliberations of Council. Article 23 dealt with the subject. Clause 1 bluntly stated: 'All project proposals shall be submitted to the Organization by members and shall be examined by the relevant committee'. Clause 4 gave Council the responsibility of arranging for the 'implementation and, with a view to ensuring their effectiveness, follow up' of projects. Clauses 5 to 7 were concerned with subject matter and criteria. Clause 8 stated that: 'The Council shall decide on the relative priorities of projects, taking into account the interests and characteristics of each of the producing regions. Initially, the Council shall give priority to

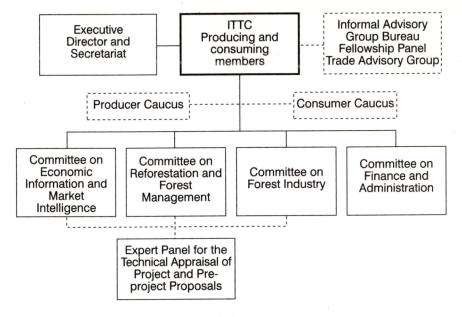

Solid lines show functions derived from the ITTA, 1994
Dotted lines show informal functions decided by Council as at May 2001

Figure 4.1 *Organizational structure and functions of ITTO*

research and development project profiles as endorsed by the Sixth Preparatory Meeting on Tropical Timber…' There was also provision for 'pre-projects', for which the rules were rather more flexible.

As mentioned in Chapter 3, the Agreement included no provision for a reforestation fund. Under Article 20, projects were to be financed from a 'special account', which could derive its funds from three sources: the Second Account of the Common Fund for Commodities; regional and international financial institutions; and voluntary contributions. The Special Account was never to prove adequate!

The prolonged 1st Session of the ITTC took place in Geneva, starting in June 1985 and ending in August 1986. It dealt with the difficult and sometimes contentious questions of where the headquarters should be located (eventually Yokohama) and who should be the Executive Director (Dr Freezailah bin Che Yeom of Malaysia was eventually elected).

Other International Initiatives

International interest and concern about the future of tropical forests was growing fast at this time. FAO declared 1985 the 'International Year of the Forest'. The World Resources Institute (WRI) convened a Task Force which prepared a report, *Tropical Forests: A Call for Action*,[1] and, stimulated by this, FAO prepared *The Tropical Forestry Action Plan*.[2] The net result was the launch of the

---

BOX 4.1 FUNCTIONS OF THE COMMITTEE ON REFORESTATION AND FOREST MANAGEMENT

'The Committee on Reforestation and Forest Management shall:

a Keep under regular review the support and assistance being provided at a national and international level for reforestation and forest management for the production of industrial tropical timber.
b Encourage the increase of technical assistance to national programmes for reforestation and forest management.
c Assess the requirements and identify all possible sources of financing for reforestation and forest management.
d Review regularly future needs of international trade in industrial tropical timber and, on this basis, identify and consider appropriate possible schemes and measures in the field of reforestation and forest management.
e Facilitate transfer of knowledge in the field of reforestation and forest management with the assistance of competent organisations.
f Coordinate and harmonise these activities for cooperation in the field of reforestation and forest management with the relevant activities pursued elsewhere, such as those under FAO, UNEP, the World Bank, regional banks and other competent organizations.'

In addition, with regard to project proposals referred to it, shall:

a 'Consider and technically appraise and evaluate project proposals.
b In accordance with general guidelines established by the Council, decide on and implement pre-project activities necessary for making recommendations on project proposals to the Council.
c Identify possible sources of finance…
d Follow up the implementation of projects and provide for the collection and dissemination of the results of projects as widely as possible for the benefit of all members.
e Make recommendation to the Council relating to projects.
f Carry out any other tasks assigned to it by the Council.'

Further:

'…each committee shall take into account the need to strengthen the training of personnel in producer member countries; to consider and propose modalities for organising or strengthening the research and development activities and capacities of members, particularly producing members; and to promote the transfer of research know-how and techniques among members, particularly among producing members.'

Source: ITTA, 1983.

Tropical Forestry Action Plan (TFAP) that year at the World Forestry Congress in Mexico City. Also in 1985, IIED completed, with the Government of Indonesia, a review of forest policies in Indonesia which, after wide consultation with Ministers, Directors of Departments, non-governmental organizations (NGOs) and others, analysed the effect of all government policies on Indonesia's forests and recommended reforms.[3]

The 2nd Session of Council – Yokohama, March 1987

Business in ITTO started in earnest at the 2nd Session, held in Yokohama. Dr Freezailah, in his introduction, made a statement that represented, in the view of many present, what should be the philosophy of ITTO; it was certainly the philosophy that he pursued as Executive Director.

> During the ten years since negotiations began to the time that ITTO became operational, …nothing significant has happened on the ground to halt the irrational rate of tropical forest depletion, or to replace the prime or noble species that have disappeared or are disappearing. If that continues, the tropical timber trade does not have much of a future. At best, it will be a very different future, much more difficult than in the past.
>
> It is in this context that the most unique role of the ITTO has to be seen. It is unique because it is the only international, intergovernmental organization, of global stature, which is solely concerned with the tropical forests, and solely focused on the production from those forests of industrial wood for export; but this production and export represents only a narrow segment of the tropical forestry problem.
>
> Nevertheless, that narrow segment is the crucial segment, and it is absolutely vital that you, members of the Council, as well as all who have gone through so much trouble to be here today, appreciate that important fact. It is in that narrow segment that a glimmer of hope for the tropical forests may be seen. Given the present conditions and attitudes, it is essential to realize that the conservation of the tropical forests will depend upon their rationally managed use…
>
> What is crystal clear is that conservation of the tropical forests is one of the specific objectives of ITTO. Indeed, it can be said that conservation is a precondition for the survival of the tropical timber trade. It is in that specific objective that the interests of conservationists, environmentalists, tropical timber producers, traders and users coincide in ITTO. It is that coincidence of interest, through which ITTO can transform its unique role into a decisive role in the interest of tropical forestry. But how is that decisive role of ITTO to be played? This is where ITTO's uniqueness comes to the fore. It is new! It is global! There is no traditional or established way to follow. No scientific formulae to be guided by. No red tape to be strangled by. The objectives of the organization are clearly before it. These objectives have to be attained, and attaining them means to balance, or try to achieve a balance between, conservation and wise utilization. ITTO's efforts can start immediately, in ways which are consistent with the urgency of the problems of the tropical timber trade and the tropical forest resource. There is nothing to dismantle, nothing to change before it can start.

A ringing declaration, indeed. No statement of purpose could be clearer. The meeting started in an up-beat mood. Attendance was good. Apart from member states, many observers were admitted: eight UN organizations, four specialized agencies of the UN, four intergovernmental organizations, and 31 NGOs, of which 14 were concerned with timber and 17 with the environment. The NGOs were made to feel welcome and participated with enthusiasm. Press coverage was vigorous. But divergent interests were already becoming apparent.

Dr Freezailah presented a draft work programme for 1987–1988, largely based on the very substantial studies carried out by the 6th Preparatory Committee, as instructed in Article 23, but modified to take account of the changed circumstances during the intervening period. These changes were an increase in domestic processing but without the expected benefits, the downturn in the market for tropical timber, and intensified competition from softwoods. He described the draft proposals for reforestation and forest management as 'constituting a programme to turn the tropical timber trade into an instrument for conservation of the tropical timber resource, rather than being a partner, however unwittingly, in its continued destruction'. He added: 'there is no viable alternative'.

Dr Freezailah intended to build ITTO upon three principles: that it should be lean, action-oriented and businesslike; that it should be capable of acting with a sense of urgency to match the speed at which tropical forests and the trade were being harmed; and that it should strengthen and complement the work of other organizations in the same field.

His introduction continued:

> From these considerations the work programme virtually sets itself. A future, which depends on the generosity, the altruism and self-sacrifice of others to conserve tropical timber forests, promises no future at all. The tropical timber trade has to generate, by and for itself, the resources to sustain its supply base. Then it will have to ensure that they are also used to that end and that end alone. One thing is certain – more of the same recipe is not the answer. For any change, present informational systems and present forest and industrial management levels are inadequate, if not inappropriate. This programme is a first step to correcting the deficiencies in an integrated way.

He then proposed projects in the fields of each of the three permanent committees.

In Economic Information and Market Intelligence, there were four projects: information on the uses to which tropical hardwoods were put; the establishment of greater market transparency about prices related to species and timber specifications; the more effective and timely dissemination of more detailed market information; and the feasibility of a forum for producers and consumers to exchange views on the world market situation and short-term prospects.

In Reforestation and Forest Management, there were six projects. The first was concerned with the conservation and management of tropical forests and focused on determining the status and dynamics of tropical forests as a basis for planning the development and utilization of the resource, and for identifying the factors contributing to tropical forest destruction, how great it was and how it was caused. The second and third were concerned with the development, management and utilization of tropical forests: the rehabilitation of the large area of logged-over forests to ensure their continued productivity through silvicultural improvement or enrichment planting; the management and utilization of primary forests, in order to harmonize management and logging such that logging operations do not result in damage to the residual trees or to the environment and impair the future productive capacity of the forests. The fourth was aimed at assembling information on the progress of forest plantations in tropical countries, including the size and characteristics of these plantations and expected outputs. The fifth would investigate questions of conservation and sustained timber supply through agri-silvicultural systems focusing on the management of high forest species for timber, while the last would examine the prospects for prime forest species such as mahogany and teak.

It was proposed that Forest Industry should concentrate on bringing lesser-known species into the market and on overcoming difficulties in the sawing, processing and seasoning of their timbers.

This, then, was the work programme that Dr Freezailah proposed for 1987–1988. But, although it was fully in line with Clause 8 of the Agreement and with a decision reached at the 1st Session of the Council, it was not received with the enthusiasm or support that it deserved. It was pointed out forcibly that, according to the Agreement, it was for members to propose projects. This was an implied rebuke to the Executive Director, which made it more difficult for him, in future, to take the lead. The representative of the US went on record as saying that 'her delegation still had serious reservations about the amount of information to judge the project proposals … and looked forward to receiving full project proposals'.[4] She later raised a number of minor objections to the budget, provoking the delegate of Papua New Guinea to remark that it seemed as though some countries did not wish the Agreement to succeed! He was backed up by an eloquent intervention by the representative of Trinidad and Tobago.

In fact, it was already apparent that the different parties had very different expectations. The main interest of the producers lay in using ITTO as a new source of funds to finance the kinds of development that they wished to see in their own countries – institutional strengthening and the widening of market opportunities. The consumers, on the other hand – who would also be the main contributors to the Special Account – wished to see their funds directed towards measures that would secure future supply and, in general, that would correspond to carefully considered priorities. The environmental NGOs, for their part, wished projects to be specially directed towards the better management and conservation of the forest resource, and the interests of local peoples. The future of ITTO was to be played out by interaction between these different groups. The dichotomy of interest between producers and consumers was undoubtedly enhanced by the adoption of the standard UNCTAD procedure whereby the Council divided from time to time into what became known as producer and consumer caucuses who met separately and took up positions which were presented to the plenary by 'spokespersons'. All observers were excluded from caucus meetings.

The Council took note of pre-project proposals to the value of US$640,000 and project outlines to the value of US$2,070,000. The executive director was authorized to proceed with the pre-projects but requested to present full project proposals for the remainder. Meanwhile, members were invited to transmit their own proposals for further projects. Work was also to begin immediately to build up a capability in market intelligence – to compile, collate and publish statistical information. This was to prove a field in which ITTO became highly effective.

The first contributions were pledged to the Special Account: US$2 million by Japan; US$1 million by Switzerland; the likelihood of US$600,000 by the Netherlands; and – a significant gesture – US$12,330 by 14 non-governmental conservation organizations. The contribution from Switzerland carried the condition, later to become very familiar – 'for the co-financing of projects approved by the Swiss authorities in accordance with their priorities'.

The session ended on an optimistic note.

The 3rd Session of Council – Yokohama, November 1987

The plenary sessions of Council meetings tend to be formal occasions which allow little opportunity for free debate; yet there were many issues in the early days of ITTO (and indeed there still are) that would benefit from a free exchange of views, involving not only the official delegates but also the many observers from NGOs, the trade and technical organizations who were flocking to Council meetings. The Council and the Executive Director considered it important to listen and consult widely. Accordingly, it was decided to hold seminars before the opening of Council meetings to encourage free discussion of some of the fundamental issues underlying the management of tropical forests for timber.

At this session, 14 of the 31 project proposals submitted were from members and 12 altogether were approved for immediate financing. Of these only three were concerned with natural forest management: forest management of natural forest in Malaysia; the biology of the Okoume (*Aucomea klaineana* Pierre) in Gabon; and investigation of the steps needed to rehabilitate the areas of East Kalimantan seriously affected by fire. Approval was also given to implement pre-projects on the study of enrichment planting and on the development of integrated approaches for sustainable utilization and conservation of tropical forest in the Amazon region.

Difficulties were already apparent in a number of fields. First, it became evident that many producer countries were having difficulty in preparing suitable proposals; even Brazil made use of three consultants to prepare the proposal for the pre-project mentioned above.

Thought was already being given to establishing some priorities; the early proposals were a distinctly mixed bag! An informal working group of the Committee on Reforestation and Forest Management prepared 'Proposed ITTO Guidelines in the Field of Reforestation and Forest Management';[5] this was noted by the Committee and reproduced in the record of the Council meeting. WWF coincidentally presented a thoughtful paper entitled 'Project concepts to promote the utilization and conservation of tropical forests'.

The Council recognized an evident need for some guide to priorities and for the formalization of what came to be known as the project cycle – the presentation, review and approval of projects. The view was already prevalent in some circles that passing projects through ITTO was a soft option compared to UNDP or the bilateral aid agencies. Accordingly, at this session, the Council requested a paper from the Secretariat on policies and guidelines for ITTO projects. This was to be the first of many steps taken to improve the presentation of projects and to refine the procedures for selecting them.

FAO was at this time preparing for a full assessment of the forest resources in 1990. This was welcomed on the grounds that it would provide a firm basis of fact about the status of the tropical forest resource. The TFAP was also launched in 1985 and its progress was watched with great interest.

It was always intended that ITTO should be a lean organization but, at this meeting, finance began to be a continuing preoccupation; the Secretariat was already sorely pressed. Many countries were defaulting on their contributions, some because of a shortage of funds, others (notably the US) for reasons of policy.

The 4th Session of Council – Rio de Janeiro, June–July 1988

The 4th Session was the first to be held in a producer country: in Rio de Janeiro in Brazil. The opening speeches set out the main preoccupations of the various parties. The Chairman, Otto Th Genee, stressed that it was 'incumbent on the ITTO to develop an overall strategy and to set priorities for action'; all states, trade organizations and other NGOs should be involved. The issue of sovereignty, which became of overwhelming importance in UNCED in 1992, was raised by the Foreign Minister of Brazil (HE Dr Roberto De Abreu Sodre). He was specific: 'The task of formulating policies to administer and safeguard natural resources remains for Brazil an exclusive sovereign responsibility of the nation. Such policies must obey the imperatives of national development priorities, in particular the pressing problem of poverty which calls for urgent solutions and international action.' The delegate of the People's Republic of Congo, for his part, 'urged the Organization to pursue policies which make it increasingly relevant to the needs of tropical timber producers'. Forests were the lasting wealth of the country; exploitation, therefore, should be rational and regulated through sound conservation practices. The concern of Côte d'Ivoire was with trade: there was a 'need for ITTO to promote equitable and remunerative terms of trade in tropical forest products and thereby provide an incentive for sustainable management and conservation of tropical forests'.

In response to the request at the 2nd Session, the Secretariat produced a paper on 'Criteria and priority areas for programme development and project work'. This gave detailed consideration to priorities for ITTO and compared them with the proposals of Preparatory Committee 6. In particular, they recommended a better balance between project and non-project work, the purpose being to shift attention towards the role of the Council as a consultative forum for the reform of policies. In accordance with the Council's formal procedures, any conclusion was presented either as a 'Decision' or a 'Resolution'. In this instance, it appeared as Resolution 1(IV), the operative part of which is given in Table 4.1. Some progress had at last been made in sorting out priorities.

Table 4.1 *Operative part of Resolution 1(IV) (July 1988)*

The International Tropical Timber Council
Decides:

1. ITTO shall concentrate its activities in the following three areas:
(a) Sustainable management of the tropical timber resource, including appropriate harvesting methods.
(b) Further processing.
(c) Market transparency and trade diversification.
2. In accordance with Art. 25, ITTO shall achieve a better balance between project and non-project activities.
3. In accordance with paragraph 1 above, the ITTO shall pursue the following activities on a priority basis.

	Non-project activities	*Project activities*
Committee on Reforestation and Forest Management	Encourage the development of national policies aimed at sustainable utilization and conservation.	Research and development and project implementation of improved forestry and reforestation practices, including harvesting methods, for the sustained production of industrial tropical timber.
Committee on Forest Industry	Monitor on-going activities and promote exchange of information and transfer of technology and encourage harmonization of nomenclature and promote the development of uniform technical specification of tropical timber products.	Activities aimed at the development of further processing, including in particular research and development to promote utilization of lesser known species.
Committee on Economic Information and Market Intelligence	Remove deficiencies in collecting, analysing, interpreting and disseminating data on trade, production and consumption, with a view to increasing market transparency.	Studies and collection of information to complement non-project work on tropical timber markets, and assistance towards strengthening the capability of producing countries in market data collection and processing.

Source: ITTC(IV)/D.1.

5

First assessment:
living in a fool's paradise

*'At present, the extent of tropical forest which is being deliberately managed
at an operational scale for the sustainable production of timber is, on a
world scale, negligible'*
A main conclusion of the ITTO/IIED study

Introduction

It soon became apparent that very little was known for certain about the actual
situation in most of the ITTO producer countries. Accordingly, in 1987, ITTO
commissioned IIED to make a study of the management of natural forest[1] for
the sustainable production of timber within the producer countries of ITTO.
This was to find out how much was being successfully managed in a sustainable
manner (so that its character as forest was preserved and its potential to
produce was maintained). Where management was succeeding, the study was
to determine what local conditions made it successful and, where it had failed,
the reasons for its failure. The study was carried out during the course of 1987
and early 1988 and was presented to the ITTC in November of that year. It
formed the core of the book *No Timber without Trees*, published by Earthscan in
1989.[2]

This study drew upon two main sources of information: country visits, and
consultation with individuals who had a detailed knowledge of the subject. A
round-table discussion was held in each continent.[3] India, although a producer
member of ITTO, unfortunately did not participate in the study. In addition,
information was included on the management of the rain forests in Queensland,
Australia, because the experience there at that time was highly relevant. Shortly
afterwards, much of the Queensland rain forest was included in a World
Heritage site and timber production ceased.

It must be emphasized that the study was concerned with management for
sustainable timber production and not with sustainable management for other
purposes, such as catchment protection or nature conservation. It was looking
for management at an operational scale, not merely demonstrations or trials.

The Definition of Issues

In order to define the status of the 'sustainable utilization and conservation' of natural forests in the countries visited and, in particular, the status of management for the sustainable production of timber, the team attempted to find answers to these general questions:

- Over what areas is natural forest managed at an operational scale for the sustainable production of timber?
- Where such management has been undertaken successfully, what are the conditions that have made this possible?
- Where such management has not proved possible, or has been attempted but failed, what have been the constraints that have made it difficult or impossible to apply?

A schedule of some more detailed questions which proved useful in analysing the situation are shown in Box 5.1. All are relevant to successful management.

The various levels at which sustainable management can be practised have been described in Chapter 2. Each can be valid in particular circumstances, but can only be considered as truly operational if a conscious decision has been made to manage the forest at that particular level, if the decision has been conscientiously applied and the results have been monitored. This implied (what was in fact the case) that there might be areas which had not been brought within any formal management structure but were not deteriorating. It would not be difficult to bring these under sustainable management.

Some other questions had to be clarified if the results of the study were not to be misunderstood.

First, it was emphasized that it was not yet possible to demonstrate conclusively that any natural tropical forest anywhere had been successfully managed for the sustainable production of timber. The reason for this is simple. This question cannot be answered with full rigour until a managed forest is in at least its third rotation, still retains the full forest structure, is fully stocked with commercial species which are growing well, and possesses adequate regeneration and an intact soil and ground flora. No tropical moist forest had been managed consistently for a sufficiently long period to fulfil all these conditions. But, as it was not particularly helpful to insist on this point, the study adopted a definition which was strict but not so stringent as to be useless in practice.

Second, whatever definition is adopted, the management status of the forest may be altered almost overnight by changes in the implementation of government policies, and this alteration may be in either direction. The best-managed forest may be arbitrarily allocated to another use – an agricultural development scheme perhaps, or even a plantation of fast-growing trees for pulpwood; or it may be decided to make a sudden and unjustifiable change in the length of the cutting cycle. At the stroke of a pen, a forest which was managed sustainably becomes one which is no longer so managed. In contrast,

BOX 5.1 QUESTIONS USED IN THE IIED STUDY

Policy. Is there a national land use policy? Is there a national policy for the sustainable management of a permanent forest estate? If not, why?

Extent. What area of natural forest is managed for the sustainable production of timber?

Allocation.
- Is there a satisfactory system for choosing, demarcating and protecting those areas that will be used as production forest? If not, why?
- Is there a satisfactory system for choosing, demarcating and protecting those areas that will be used as protection/conservation forest? If not, why?
- Are there pressures from other sectors or interests to remove productive forest from forest use? What measures are being taken to counter or divert these pressures?

Sociological and economic conditions. In what ways do the various people who have an interest in or are affected by the management of the forest, benefit from this management or suffer from mismanagement (people dwelling in or near the forest, loggers, middlemen, wood processors, small industries, the Forest Authority, consumers generally, other government revenues)? Are the benefits adequate to provide an incentive to good management? Is there equitable distribution of these benefits? If not, why?

Management. Are the objectives of management conducive to sustainable production? Are the management prescriptions appropriate for the particular forest type? Are they rigorously applied and reviewed? If not, why?

Pre-exploitation survey. How comprehensive and adequate is the pre-exploitation survey: choice and marking of trees for felling; analysis of trees to remain unfelled; existing regeneration; environmental conditions; routing of extraction roads? If inadequate, why is this so?

Choice of exploiters. Does the choice take into account the best long-term interests of the forest? How?

Conditions of exploitation.
- Do these bring reasonable benefits to the various parties concerned: government revenues, any reforestation fund, the logging companies, local contractors, logging labour, those with customary rights in the land?
- Are the conditions of exploitation such as to encourage long-term investment in the sustainable management of the forest? Are there reasonable incentives to encourage good management? What proportion of revenues are returned to forest management? If these conditions are not met, what prevents it?

Quality of exploitation.
- Are there guidelines for the siting, construction and maintenance of extraction roads, weather in which exploitation should not take place, equipment to be used, directional felling, cutting of lianes, etc? Are such guidelines followed? If not, why?
- Are the above conditions monitored during and after exploitation? How? How well?

Post-exploitation survey and treatment. Are there guidelines? Are they sensitive to different forest types? Are they adhered to? Is later performance monitored? How? If not, why?

Control. Is there effective control of operations at all stages? If not, why?

Follow-up. Are there arrangements for monitoring and reviewing prescriptions? If not, why?

Research. Is research designed to support sustainable timber production from natural forest? Is it adequate to provide the necessary information to answer the questions set out above? Are there permanent sample plots to provide the data upon which sustainable yield can be calculated? Are the data processed and made available to management within a reasonable time?

Education and training. Are sufficient trained staff at all levels being produced with qualifications in the skills needed in natural forest management?

Source: Poore et al, 1989.

a forest which is at one time outside state control, may later be demarcated, provided with a management plan and given effective control. It then can be considered to have a reasonable chance of being managed for sustainable production in the future. The figures given in the study represented the best judgement that could be made about the prospects of areas under different forms of management. Changes in the implementation of policies could have readily led to them being revised substantially – either upwards or downwards.

The report stressed that it was vital that governments should fully understand this. Management that was good could easily be turned into management that was destructive; but, on the other hand, the substantial areas of forest that were now very nearly under a sustainable management system could readily be turned into forests which were being managed in a satisfactory and convincing manner for sustainable production.

Accordingly, the team adopted a strict definition of forest management – that it should be practised on an operational rather than an experimental scale and that it should include the essential tools of management, these being: objectives, felling cycles, working plans, yield control and prediction, sample plots, protection, logging concessions, short-term forest licences, roads, boundaries, costings, annual records and the organization of silvicultural work. They also had to meet wider political, social and economic criteria without which sustainability was likely to be unattainable.

Some considered that this definition was too strict; others thought it too lenient. Being criticized from both sides may mean that the team might have got it about right! If a more relaxed definition had been used, it would have bred attitudes of dangerous complacency. Instead, what was needed was to generate a sense of great urgency but of qualified optimism.

The results of the study, using this strict analysis, provided a picture that was not encouraging, but there was also a brighter side. There were many areas where *some* of the elements of management were in place; many silvicultural trials had produced encouraging results, and there were many signs that countries were taking some of the steps needed to establish sustainable management at an operational scale.

Box 5.2 Examples of forests managed at different levels of intensity

These examples are chosen from forests that are managed at very different levels of intensity. In all of them some elements of sustainability are present; but in none of them has sustainable management been convincingly demonstrated in practice. This might be possible, however, with some relatively small additional effort.

Wait and see. This policy was being deliberately applied in the Tapajos National Forest in Brazil. The forest was reserved; knowledge was available upon which to manage it; but operational management had not begun because there was not yet a sufficiently profitable market, a situation caused by the abundant supply of very cheap timber coming from land being cleared for agriculture.

Log and leave. Forests treated in this way covered very large areas in all continents. This would be an acceptable management regime if it were the result of a deliberate management decision based on reliable information about potential future crops and if it were adequately controlled. But unfortunately this was seldom the case. Frequently there was no proper security of tenure and no determination of logging standards or of the period to rest before re-logging. Such forests were very vulnerable to abuse. In many African forests, logging was very light and the enhancement of growth or regeneration through logging was slight; such forests, if left without apparent attention from forest departments, were a great temptation to colonists. In other areas, logging might be too heavy or there were strong inducements to enter the forest to re-log for other species or stems of lower girth and quality. It was considered that both increased timber prices and the entry of 'lesser-known species' into the market could, without adequate control, have had an adverse effect.

Minimum intervention. Such a system was being deliberately practised in Queensland. The possibilities of both stand improvement and enrichment planting were rejected on economic grounds. The resulting lack of intervention had considerable environmental advantages in preserving biological diversity. It was considered possible that much more forest might become managed in this way in the future, as there seemed to be a prevailing trend towards polycyclic systems and a recognition that stand improvement had often led to the elimination of species that later became marketable. The environmental benefits might also become better appreciated.

Stand improvement and enrichment planting. There were many examples, but almost all had failed so far to proceed beyond the project scale.

Source: Poore et al, 1989.

A few examples are given in Box 5.2 that illustrate how important elements of sustainability proved possible in forests managed at very different levels of intensity, although, in none of these cases, had sustainable management yet been convincingly demonstrated in practice.

The Findings of the Study

The results of this country by country assessment were not encouraging.

In Latin America and the Caribbean, the total area being sustainably managed at an operational level was limited to 75,000ha in Trinidad and Tobago, of which 16,000 had been 'declared as fully regenerated after logging'. Management was not intensive, but adequate, although silvicultural treatments were rarely applied and management-plan prescriptions were not strictly followed. There had been some striking recent advances in other countries in the region, but there was a lack of advance planning of the location and intensity of the annual cut; supervision and control were weak; and there was little protection against unplanned settlement and uncontrolled logging.

The conclusion for the African countries was similar. No sustained yield management over large areas was being practised; management had been progressively abandoned, possibly with the partial exception of Ghana. A selection system had been tested experimentally in Côte d'Ivoire for eight years which would be applied to 10,000ha of Yapo Forest.

The situation in Asia was more complicated. It was not possible to give an estimate of the area of forest genuinely under sustained yield management. But, with the exception of Papua New Guinea, all the forests under concession agreements within the region were, at least nominally, under management. But there was a very great difference between theory and practice in many parts of the region. Management tended to be deficient in many features: the adequacy and protection of the residual stand; the amount of silvicultural work following the felling cycle; the prohibition of re-logging between cycles; maintenance of roads and control of post-felling erosion; the protection of unworked forests; and the writing and enforcement of working plans. This difference between theory and practice could not be quantified without much more detailed inspection than was possible in the available time.

The most encouraging and complete theoretical system for operational management in the region was in place in Peninsular Malaysia, where the so-called 'selective management system' provided for stand modelling based on adequate inventory and the data from permanent sample plots. The system had only recently come into full use and there was no certainty of its success into and after the first cutting cycle, but it contained many of the elements necessary to respond flexibly to the condition of the forest stand. It was intended to apply this system to the total reserved production forest within the legally constituted Permanent Forest Estate of Peninsular Malaysia. Some forests and operations appeared to be reasonably successful as sustained yield units: one in Thailand, several in Peninsular Malaysia, and a small selection in Sabah, the Philippines and Indonesia.

The conclusion was striking: the area of tropical moist forest demonstrably under sustained yield management for timber production in the producer countries of ITTO (with the exception of India) was, at the very most, about 800,000ha – this out of an estimated total area, in 1985, of some 828 million hectares of productive tropical forest remaining in all tropical forest countries.[4]

Clearly, urgent action was required, not only to ensure the proper management of previously unlogged forest, but also to assess the status of logged forest and degraded forest lands, to plan remedial action and to bring these into sustainable production as rapidly as possible.

The Conditions for Sustainable Production

An important part of the study and much of the attention of the round-table discussions held in each of the continents was devoted to identifying the conditions under which management for sustainable timber production had hitherto been successful. There were four: long-term security, operational control, a suitable financial environment and adequate information. If one were lacking, sustainable timber production would fail. The report produced arguments for each of these.

Long-term security

The first essential was the existence of a PFE for timber production. Management for sustainable production was only possible if the forest continued to exist; without forest there could be no production. The study found that the lack of any guarantee that forest would remain as forest proved to be the overriding reason for failure. Moreover, the absence of any guarantee of security of tenure disinclined the forest manager, whoever he might be, to invest time and money in future management. Examples of this from Peninsular Malaysia, Ecuador and Queensland are given in Box 5.3.

In each of these instances, central government had not had a firm and consistent policy with regard to maintaining a permanent forest estate for timber production. This meant that it had either changed its policy or had willed insufficient resources to defend its policy. In the case of Queensland, a conflict between state and federal interests was also involved.

From these experiences it was apparent that governments of producer countries should consider it a matter of high priority and importance to set aside, as a part of an overall land use policy, a permanent forest estate which it should be prepared to defend both practically and politically. It was, of course, impossible for governments to identify in advance all claims that might arise in the future for alienating parts of this estate, but there were a number of general trends that could not be ignored. It would be a self-defeating policy, for example, to include within the permanent productive forest estate extensive areas that were highly suitable for the sustainable production of food or for profitable non-food cash crops. There would also, almost certainly, be increasing pressure from an environmentally sensitive public to make sure that national land use plans made adequate provision for the protection of catchments, for the rights of indigenous peoples and for the conservation of ecosystems and plant and animal species; also, that management of the forest for timber production was carried out in a sustainable manner and in such a way that it did not cause environmental damage. Experience had shown that the growth of

---

BOX 5.3 EXAMPLES OF FORESTS WITHOUT LONG-TERM SECURITY

In these examples from Peninsular Malaysia, Ecuador and Queensland, sustainable forest management became impossible because there was no long-term security for the forest estate.

In Peninsular Malaysia, there had been developed one of the potentially most effective silvicultural systems, the Malayan Uniform System, to manage the very productive lowland dipterocarp forests of the peninsula. Early results with this system were most promising. But the forests to which it was applied were situated on soils which were eminently suitable for agricultural cash crops and, as a matter of government policy, appropriate stands were alienated after the first cutting cycle for conversion to plantations of rubber and oil palm.

In Ecuador, a number of logging concessions, which were operating until 1970–1980 in defined areas with defined yields, were cancelled because of 'the practical and political impossibility of protecting the concession areas from occupation and cultivation by colonists'. As a result of this experience, the government effectively abandoned the use of concession agreements as a tool of forest management.

In Queensland, Australia, a system of Permanent Forest Reserves had been defined by legislation and the boundaries marked on the ground; this was part of a land use policy which also allocated a substantial fraction of the forest area for nature conservation and for catchment protection. A proportion of the area so reserved was allocated for the harvesting of timber and a system of 'minimum intervention' logging was devised, which was the most complete example of sustainable management that the team found anywhere in the tropics. Yet, because the whole of this area became a World Heritage Site, timber harvesting was suspended by the fiat of the federal government and has never resumed.

In two of these cases, a sector of the population deemed it more important to replace the use of the forest for timber production with some other use – in fact the policy had not carried an influential section of the population along with it. In the case of Ecuador the forest had been invaded by colonists who saw greater advantage in converting it to farming land; in the case of Queensland the opposition to logging came from environmentalists.

Source: Poore et al, 1989.

---

environmental awareness was an inevitable concomitant of development. There were also likely to be claims for additional physical developments: roads, towns, reservoirs, etc. Because of the need for permanence in the forest estate, any claims of this sort should be subject to the very strictest scrutiny.

The arguments needed to justify the size and location of the permanent productive forest estate against the claims of other land users would have to be politically cogent and convincingly argued. Three parties would have to be convinced: the economic and planning ministries of governments – that the goods and services to be derived from national forests and forest lands were of high importance for the future well-being of the country as a whole and to parts of the whole; local populations – that they could derive greater benefit from well-sited national forests than from alternative forms of land use; and the environmentalists – that there would be benefit in managing some forests for production and that this management could be carried out in a sustainable and environmentally acceptable way.

Security for the managers

Forest management is a long-term enterprise. The managers, therefore, would also require security. These might be government agencies – this was at present the most normal in producer countries – but there are other possible models. Forests could be managed sustainably by other agents if the right conditions were provided. Security of tenure, though essential, was not enough to ensure that forests were well managed; the three elements of security, self-interest and incentive seemed to be necessary to ensure sustainability.

Secure ownership by forest management companies, under clearly defined government conditions that the forest must remain as 'natural forest', might therefore provide a satisfactory alternative to state ownership and management. Another possible model was the leasing of the land to local communities for management under defined conditions.

Operational control

The significant conditions for operational control were: clearly defined management objectives for the forest and a plan of management; standards and defined procedures; adequate control of harvesting based upon both of these; and an assessment procedure to determine the initial average length of the cutting cycle and any subsequent refinements of it.

After the security of the forest, the most important condition for sustainable production was the control of the whole harvesting operation. There were two reasons for this: the regulation of the future crop and the reduction of environmental side effects. The former was mainly concerned with the quality of residuals and advance growth, the presence or absence of regeneration, and (possibly) the maintenance of site fertility; the latter was concerned with the dangers of soil erosion and the possible reduction of the variety of species and of genotypes.

As far as the future crop was concerned, management should remain flexible; the behaviour of the forest and of the market could be predicted, but imperfectly. There were obvious merits in a flexible predictive model for management which operated through the setting of standards and through control of such matters as girth limits for felling, number of residuals and composition of residual stand, regulations about leaving or removing damaged trees, species of lesser value or potential seed trees, differential stumpages, and many other features. All of these could be manipulated through the control of harvesting.

Much less was known about the effect of harvesting and subsequent management on 'biological diversity'. But, as information became available, it could be added to any such dynamic model so that, in theory at least, environmental advantages could be set against additional costs or reduced harvestable yield. The same considerations applied to other, non-timber forest products.

The most evident environmental effect of logging was, however, erosion. This could be prevented by the operation of rigorous standards for the design and control of roads and the use of equipment. Benefits, costs and practicability would of course enter into the choice of these; but in many cases better road systems and machinery would lead to the possibility of extraction at lower cost.

It was emphasized that standards, control and prediction all depended absolutely on adequate information of the right amount and quality, and a reliable system of control – whether operated by government, private owners or communities; the more sophisticated the management, the greater the requirements for information.

It was difficult to say whether lack of security or lack of control was the most common reason for the failure of sustainable production, but it was unfortunately true that lack of adequate control was almost universal, outside those rare examples quoted above where operational management was succeeding – and even to some extent within them. This failure could take many forms: not allowing adequate recovery time after harvesting; the marking of trees for felling not being done or being negligently done; undue logging damage; insufficient residuals; insufficient regeneration; re-logging too early; illegal logging; concessions given to unsatisfactory operators or given for the wrong reasons; failure to exact penalties for faulty logging practice; and many others.

As the vast majority of operations were ultimately controlled either by government departments or government agencies, the failure of control must ultimately be laid at the door of governments. The symptoms were staff shortages or low staff morale; poor conditions; insufficient equipment; inadequate field supervision; inadequate or inappropriate training; and insufficient research information. In an extreme form this could mean that the lives of staff were endangered if they showed themselves to be too conscientious.

All of these findings were symptoms of the low ranking of forestry in the priorities of governments and also in the view of many influential citizens, all of whom tended to view the forests as a resource to be mined, as constituting a residual use and as less valuable once exploited than alternatives uses. Many of the recommended prescriptions were concerned with these issues.

A suitable financial environment

The third condition required for sustainable production was a market for the produce. The absence of a market was the main reason that large areas of forests in the Central African countries remained unexploited; the low volume of timber removed from them was cited as one of the principal causes of the absence of adequate regeneration within them. The management plan for the Tapajos forest in Brazil had been put on ice because there was not yet a sufficiently lucrative market for its timber. On the other hand, the flexible and successful management for stand improvement in Trinidad had largely been made possible because a local market for charcoal had been skilfully used by the forest managers to carry out silvicultural operations at low cost.

It was argued that the presence of a market by no means ensured that management would be sustainable. There were unfortunately too many examples to show that the opposite was usually the case; there was a critical difference between self-interest in exploitation and self-interest in sustained management. A market was a necessary, but not a sufficient, condition.

There should also be a direct relationship between the market and the degree of intensity of management that proves profitable. The management of natural forest had one potential advantage over intensive plantations – it was easier to play the market if the systems of management were not too rigid (this was what was done in effect in the long-standing forests of Europe). It would not be easy to do so, however, as long as the market was partially supplied by the first cut from virgin forest and from lands which were being cleared for agriculture at the same time as from forests under effective sustained yield management – and in direct competition with them.[5] In fact, forestry suffered from operating within three distinctly different economic regimes: mining; sustained yield management; and the 'agricultural'. It should exploit the strengths of natural forest management – its flexibility, the quality of its products and the comparative advantage afforded by products that were well known and established in the trade. The alleged disadvantage – the long time of return – was mainly evident when the whole operation of a concession or a government forest was not turning over on a sustainable rotation, and if some of the initial profits derived from the first cut had not been reinvested in management.

Adequate information

The last prerequisite was adequate relevant information for policy and management, carefully chosen. Immense amounts of energy and resources could be wasted in pursuing the wrong information or gathering data at the wrong level of precision. There were critical deficiencies in knowledge in most countries about the resource, likely markets and the information required for management.

National policy required reliable information on the extent of the forest estate; the present commercial volume; the growth potential of the forest; and the nature and extent of other claims on forest land (to honour the traditional rights of forest peoples, for the growth of food and cash crops, for catchment protection, for conservation, for mineral exploitation, and for reservoirs, roads and human settlements). In order to have enough information for management and control, for example, inventories were required both for harvesting and for management, diagnostic sampling for regeneration, the results of measurements from permanent sample plots, and knowledge of the ecology of the principal economic species. There was a need also for accurate financial figures for the costs and returns from the whole range of activities associated with forest management, and with the harvesting and marketing of forest products.

To obtain all this information requires a timely and well-planned investment in survey, research and monitoring. Without it, it would not be possible to exercise proper judgement in the selection of the permanent forest estate, nor to provide a favourable financial environment for natural forest management, nor yet to establish the standards and criteria for controlling the forest operation.

Conditions of Success

The conditions of success could be clearly identified as the following: government resolve to set aside a forest estate for the production of timber and to manage it sustainably; a sound political case for the selection of a permanent forest estate as part of a national land use policy; guaranteed security for the forest estate, once chosen; an assured and stable market for forest produce; adequate information for the selection of the forest estate and for planning and controlling its management; a flexible predictive system for planning and control based on reliable information about growth and yield; the resources and conditions needed for control; and the will needed by all concerned in order to accomplish effective control.[6]

Recommendations to the International Tropical Timber Council – November 1988

The team presented its recommendations for action to the ITTC in November 1988. From the time that IIED received this commission, it had been its intention that this report should stimulate the Council to develop an Action Plan, and the recommendations were formulated with this in view.

The following elements were suggested as necessary for establishing on a firm, operational base the management of natural forests for the sustainable production of timber:

- Inspiring governments with a sense of urgency and purpose.
- The establishment in producer countries of a permanent forest estate, whose future is guaranteed, as part of a land use plan which also provides adequately for protection and conservation forests.
- The development of the intellectual basis and justification for tropical forest management for the sustainable production of timber at various different levels of intensity (as explained in Chapter 2). This would include consideration of the economic, social, institutional and environmental context of management at each level.
- The wide promotion of such approaches in producer countries.
- The establishment of standards and manuals of 'best practice' for all the elements of management, harvesting and supporting research.
- The encouragement of proper control through adherence to these standards and practices.
- Promoting in consumer countries an appreciation of the scale and complexity of the problem.

- Working through both producers and consumers to develop the best financial environment for sustainable production.
- Encouraging and financing the establishment of different models of management wherever the political environment is favourable. These models would be chosen so that their results might be of wide applicability or demonstrate innovative approaches.
- Encouraging an institutional environment favourable to consistent, long-term management for sustainable production by developing and encouraging appropriate training, policy guidelines, management structures, predictive modelling both of supply and demand, and research and monitoring systems.
- Encouraging cooperation between the environmental movement and the trade to create the conditions for sustainable management of properly chosen forests for production.

Not all these actions needed to be carried out by ITTO itself; indeed, the Agreement specified that ITTO should, wherever possible, use the resources of other organizations in the field. Many of these actions, for example, were appropriate both to ITTO and to FAO, and it was clearly the responsibility of those governments which participated in the governance of both organizations to decide what the balance should be.

The report continued:

> But there are certain kinds of action in which ITTO has a clear responsibility as a treaty organization of producers and consumers. These are issues of tropical forest management which are concerned with the promotion, harmonization and sustention of the trade flow in tropical timbers and the sustainable management of the forests upon which this trade depends... In this connection, we would argue that all the issues that we have raised are sufficiently important for the future of the tropical timber trade for ITTO, even if it does not implement them itself, to play an active part in inducing others to do so!

Recommended actions

The action needed could be divided logically into four groups: actions of *promotion* – those aimed at persuading all those engaged in natural forest management or affected by it; actions of *diagnosis* – concerned with the further examination of critical issues to determine exactly where action was most needed and what that action should be; actions concerned with *providing examples* – to give high publicity to existing models of success and to develop new models; and *facilitation* – to provide any necessary aids to make the large-scale expansion of successful management easier, more rapid and more effective.

Action was recommended under the following headings:

- *Promotion*
  Policy reviews embracing the forestry sector
  Study of the amount of tropical forest that must be retained
  The economic case for natural forest management
  ITTO-sponsored meetings
- *Diagnosis*
  Tropical forest resource assessments (natural forests and plantations)
  Tropical timbers: financial aspects of harvesting, management and trade
  Identification of field projects where sustainable production of timber
  and non-timber products may be combined
  Feasibility of measures to encourage sustainable management
- *Examples*
  Innovative models of management
- *Facilitation*
  Guidelines of best practice
  Guidelines for a minimum research design
  Permanent sample plots
  Appropriate training
  Tropical forest management information network
  Exchange visits.

Consideration by the International Tropical Timber Council at its 5th Session

Special arrangements were made for the consideration of this report at the International Seminar on Sustainable Utilization and Conservation of Tropical Forests held on 12 November 1988, just before the Council Session. This was very well attended. Participants could have heard a pin drop when slides were shown of the area of forest under demonstrated sustainable management for timber production as compared with the total area of tropical forest! But no delegate disputed the findings. Although the Executive Director referred to the report in his opening statement, there is no record of any formal discussion of the report in the Council meeting. However, the Council 'endorsed the recommendations by the Working Group to the Permanent Committee on Reforestation and Forest Management ... regarding the development of action programmes for project and non-project work based on pre-project studies carried out by IIED, HIID and JOFCA.'[7]

The seed was sown. There was a changed perception of the scale of the problem; a sense of urgency was generated; and, gradually, some of the most important recommended actions were realized.

6

From Abidjan to Bali:
a radical new agenda

'The tropical timber producing countries fully support the set target of the year 2000'
The Representative of Brazil as Spokesman
for the Producer Members

An important international development which coloured events during the period between the Council meetings in Abidjan and Bali was the publication in 1987 of the report of the World Commission on Environment and Development (the 'Brundtland Report').[1] It set out to show that sustainable development[2] was possible and was a proper aim for world society. After Brundtland, sustainable development became a respectable aim of international policy! Previously, it had been the pursuit of a minority.

The Context

The release of the ITTO/IIED study was followed so rapidly by a number of developments within ITTO that it is hard not to draw the conclusion that they were cause and effect. More details will be given about these later in this chapter.

Meanwhile, ITTO had commissioned another study from the Harvard Institute for International Development (HIID) to review the state of knowledge on multiple-use management of tropical hardwood forests and the potential role that non-timber forest products and services can play in the sustainment of the forests.[3] This dealt comprehensively and persuasively with the economic and political case, based on good science, for the sustainable management of tropical forests for the services they provide and for their non-wood products. It dealt in turn with: tropical forest resources and the timber trade; natural forest management; the undervaluation of tropical timber; non-timber forest products and environmental services – major components of forest value; the economics of multiple-use management; silviculture and technology for multiple-use management; plantation forestry; conservation of genetic resources; customary rights versus state ownership; government policies; and international cooperation. It concluded:

> Multiple-use management of tropical forests is but one potential
> solution to the problem of disappearing forest resources… It is not by
> timber alone that forests can generate substantial returns, yield social
> benefits, and enhance development in those countries that harbor
> them… But mixed-use management, in isolation, cannot halt the
> accelerating disappearance of tropical forests worldwide… As the
> benefits flowing from forests are both local and global in scale, so must
> the costs of sustaining them be borne both locally and globally. In the
> final analysis, the benefits from sustaining tropical forests for both
> current and future generations far surpass the costs.

Although the ITTO/HIID study did not apparently lead to many direct actions,
it had an important underlying effect on the philosophy and approach of ITTO.
This can be traced in the portfolio of projects in the following years.

An important new issue had appeared on the agenda in Yokohama in
November 1988 – one with possible serious consequences for the trade. This
was the proposal for a boycott by the consumer countries on the import of
tropical timber. This seems to have arisen first in the deliberations of the
European parliament and was stimulated by two widespread beliefs: first, that
timber extraction was an important cause of forest destruction and degradation;
and, second, that a boycott on the import of tropical timber by the industrialized
west would significantly reduce this damage. These points will be discussed in
more detail later. In his introductory address to the Council, the Executive
Director made a strong appeal to all members to disavow such a boycott. At the
same meeting, there was a proposal from the European timber trade associations
to provide ITTO with a management and development fund through a levy or
surcharge on imported tropical timber. The producers naturally viewed both
proposals with some concern – the boycott because it would altogether deny
them an export market, and the surcharge because it would push up prices and
make tropical timber less competitive. These developments affecting trade issues
eventually converged in the twin movements – that timber should come from
sustainably managed sources (the Year 2000 Objective), and certification.

At this juncture, 50 per cent of all ITTO funds allocated so far had gone
into projects on aspects of reforestation and forest management; and now two
significant new projects were approved: into the conservation status of species
whose timber entered the trade, and the second studying conservation concerns
relating to the diversification of species extracted for timber (ie the effect of
concentrating on extracting and marketing the 'lesser-known species').

The 6th Session of Council – Abidjan, May 1989

The 6th Session of the Council, held in Abidjan, saw a number of important
developments.[4] The preparation of an Action Plan gathered momentum. A draft
had already been developed by the Committee on Reforestation and Forest
Management and there were calls for action plans to be prepared by the other
two committees (Forest Industry, and Economic Information and Market

Intelligence), the intention being that the three should be combined into a single action plan for ITTO. The pressure for this came mainly from the consumers – also the contributors to the Special Account – who wished to see that the funds were disbursed according to agreed priorities. They were supported by the NGOs. Another major innovation at this session was the inauguration of a Market Discussion by a joint session of the three committees. This lasted two days and was considered so successful in illuminating issues in the tropical timber market that it became a regular item on the agenda of future Council sessions.

There had been some movement in relation to the threat of a boycott. The Union Commerciale des Bois Tropicaux (UCBT) had endorsed the idea of a levy and the Malaysian Government, particularly concerned about the adverse publicity that they were receiving for the alleged rape of their forests and mistreatment of the native peoples of Sarawak, sent a high-level mission under the leadership of the Federal Minister for Primary Industry to visit the capitals of Europe and explain their position.[5] For some time, there had been criticism of the effects of the forest policies and practices of some nations, notably Brazil and Malaysia, on their forest-dwelling and forest-dependent peoples, and this was brought to a head by outspoken criticism from the Swiss citizen, Bruno Manser, over the plight of the Penan people in Sarawak. It was at this stage that the question of indigenous peoples became a living issue in the deliberations of ITTO. Previously, the main environmental concerns had been in relation to the degradation of forest ecosystems and the loss of biological diversity. But, from the 7th Session onwards, Survival International joined the ranks of the more regular observers at the meetings of the Council.

It was undoubtedly a combination of this Malaysian concern and the able diplomacy of the Executive Director that led to an unprecedented and highly significant *démarche* by the Federal Government of Malaysia and the State Government of Sarawak – an invitation to send an official ITTO Mission to investigate the situation in Sarawak and to make recommendations. The Chief Minister of Sarawak attended the Session in Abidjan in person to explain the nature of his proposal.

In his address, he emphasized that forestry was a significant contributor to the economic development of Sarawak. He referred to certain criticisms levelled against forest development in the State and took issue with those who advocated sanctions against Sarawak. Such actions, he argued, would be counter-productive; they ran counter to the ultimate objective of ensuring human survival through rational development and utilization of tropical forest resources and their conservation. Timber should be correctly viewed both as an economic factor and as a factor having a great impact on the environment. In most tropical countries, governments were unable to control forest exploitation and they needed resources if they were to reinforce conservation practices. But the incentive for conservation would only be possible if the future of timber could be assured as an economic factor in development, just as the presently developed countries had realized in the past during their development. The call for conservation should not be selectively focused on tropical timbers. It should extend to timbers all over the world. The Chief Minister stressed that developing countries were the first to be concerned with the conservation of their own forest resource. The policies to address these problems had, of necessity, to be evolved gradually and, with time, made to constitute a 'living practice'. He called upon ITTO to assist countries in their efforts at sustainable forest management. The environment and forest conservation were matters of global concern and responsibility, but should be seen in a global context with due regard to national sovereignty and sensibilities. For this reason, he said, the Government of Malaysia was ready to welcome a mission from ITTO to visit Sarawak to assist in promoting sustainable forest management. By acting in such a way, ITTO would be setting the stage for the kind of international cooperation which would enable producing countries to develop minimum standards of silviculture with the help of the developed, consuming countries. This would dispel any negative reaction and frustration on the part of producers, consumers and environmentalists.

This invitation led, as might have been expected, to vigorous debate, and the resulting resolution was, inevitably, a compromise.[6] The NGOs who attended played a significant part. The Executive Director had issued over 100 invitations to conservation bodies, the timber trade and industry. Only 15 attended, but their influence was out of proportion to their number. Apart from the usual environmental concerns, they were most particularly anxious that the terms of reference of the Mission would enable it to investigate the circumstance surrounding the traditional rights of the native peoples of Sarawak. They were at least partially successful, in that this was not specifically excluded from the terms of reference.[7]

The deliberations of the producer and consumer caucuses were, as usual, conducted in private and do not appear in the record of the meeting, but some sense of them can be gained from the final statements made by delegates.

The Representative of Malaysia:

> paid tribute to the initiatives undertaken by the Executive Director and referred to the Statement made to the Council by the Chief Minister of Sarawak, Malaysia. The Government of Malaysia, in exercise of its sovereign will, had decided to respond to the concern expressed in several quarters over the forestry situation in Sarawak. In an act of international cooperation, it had offered to receive a mission of the ITTO which would make an independent assessment of the forestry situation in Sarawak in order to assist the authorities in promoting sustainable forest management. Malaysia would respect the independence of the mission, assist in facilitating its contacts, meetings and visits, and provide information needed for the work. Malaysia would equally appeal to all to respect the independence of the mission and desist from any action that would hamper the work of the mission. Malaysia had undertaken consultations in the Council aimed at achieving consensus, if not a large measure of agreement, over the question. The draft Resolution placed before the Council was the outcome of intensive negotiations and discussions.

The most significant objection came from the Brazilian Delegation which questioned the competence of ITTO to carry out such a Mission. After the adoption of the resolution (see Box 6.1), the Representative of Brazil, on the

Box 6.1

RESOLUTION 1(VI)
THE PROMOTION OF SUSTAINABLE FOREST MANAGEMENT
A CASE STUDY IN SARAWAK, MALAYSIA

The International Tropical Timber Council,

Reaffirming the obligation and commitment of all Members to the objectives of the ITTA, 1983,

Bearing in mind Article 1(a) of the ITTA, 'to provide an effective framework for cooperation and consultation between tropical timber producing and consuming Members with regard to all relevant parts of the tropical timber economy', and Article 1(h) of the ITTA, 'to encourage the development of national policies aimed at sustainable utilization and conservation of tropical forests and their genetic resources, and at maintaining the ecological balance in the regions concerned',

Recalling the statement made by the Representative of Malaysia at the Fifth Session of the International Tropical Timber Council informing the Council of the serious efforts to promote sustainable forest management in Malaysia and inviting international assistance to support the implementation of these policies,

Taking note of the Statement made by the representative of Malaysia at its current Session of the ITTC,

Expressing its appreciation to the Government of Malaysia for its readiness to welcome a Mission to visit Sarawak, Malaysia at a date to be decided by mutual agreement,

1   Establishes a Mission with the following terms of reference:
    a   To assess the sustainable utilization and conservation of tropical forests and their genetic resources as well as the maintenance of ecological balance in Sarawak, Malaysia, taking fully into account the need for proper and effective conservation and development of tropical timber forests with a view to ensuring their optimum utilization while maintaining the ecological balance, in the light of recent ITTO studies on forest management for sustainable timber production in Member countries and relevant reports by other organizations;
    b   Based on their findings, to make recommendations for further strengthening of sustainable forest management policies and practices, including areas of international cooperation and assistance.
2   Authorizes financing not exceeding $330,000 from the Pre-Project Sub-Account for the work of the Mission.
3   Appreciates the readiness of the Government of Malaysia to fully cooperate in facilitating the work of the Mission and to allow it to visit any part of Sarawak, to meet any person and also to make available information relevant to the work of the Mission.
4   Invites all Members, and relevant international organizations and international institutions, to lend their fullest support for the success of the Mission.
5   Appeals to all Members, bearing in mind Article 30 of the ITTA, to use their best endeavours to cooperate to promote the attainment of the objectives of the ITTA and avoid any action contrary thereto.
6   Requests the Executive Director to take all necessary measures for the implementation of this Resolution and to prepare the necessary documentation for this purpose.
7   Requests the Executive Director to communicate this Resolution to all international organizations and others interested in the work of the ITTO.

Further requests the Mission to present, on a confidential basis, a Progress Report at the Seventh Session and its final Report at the Eighth Session.

Source: ITTC(VI)/D.1.

instructions of his Government, made a declaration to place Brazil's reservations on record. The Government of Brazil believed:

> that it was a matter within the sovereign discretion of any Member State to request or welcome a mission to examine and evaluate its forest resources and formulate recommendations on forest management. The ITTA made provision for necessary means and procedures to allow States to receive from the ITTO any cooperation they decided on. Brazil considered the resolution as 'sui generis', a unique and exceptional initiative outside the competence of the ITTO which [was] a commodity organization. Brazil did not recognize the competence of ITTO, through a resolution, to send a mission to a Member State or to establish investigatory procedures on national matters. The Resolution adopted by the Council should not constitute a precedent to be accepted by the Members.

The Representative of Peru also expressed his Government's reservations:

> Peru recognized the sovereign right of any State to take its own decisions. Peru would not interfere with the exercise of such rights. The ITTA, however, made no provision for sending a mission, if not so requested, to Member States. The Resolution should be seen as unique and should not constitute a precedent in the Organization.

The Representative of Malaysia responded by reiterating that his Government's decision on the matter had been taken in exercise of its sovereign discretion. Malaysia also recognized that no mission could be sent into any State without the receiving State's consent. The adoption of the terms of the resolution testified clearly to ITTO's competence in the question. He assured the Council 'that the initiative taken by Malaysia was not in any way predicated on establishing a precedent for the ITTO'. In the course of his speech, he remarked that if any State were to place any impediment in the way of the Mission, this would be an infringement of sovereignty.

Behind the turgid wording of the resolution (characteristic of contentious negotiated texts), there were, indeed, 'intensive negotiations and discussions'. Apart from the question of competence, there were concerns about the genuine independence of the Mission and whether or not it could be given powers to examine those social questions that so exercised the NGOs and the media. The NGOs, for their part, lobbied intensively to try to ensure that the Mission would be enabled to investigate and make recommendations on native land rights. There was much discussion, too, about the composition of the Mission.

Meanwhile, progress was being made in the Tropical Forestry Action Plan (TFAP). Jean Clement, on its behalf, announced that 52 countries, including 14 ITTO members, had conducted country reviews and that six ITTO members had drawn up plans of action. The NGOs continued to influence the Council in the selection of suitable projects; Theo Anderson, of Friends of the Earth (FoE) – Ghana pledged US$14,000, partly to be used for developing a project on the 'identification and promotion of sustainable management of natural forests for timber production'.

The 7th Session of Council – Yokohama, November 1989

The 7th Session took place against a background of increasing interest in tropical forests. The Executive Director drew attention to the way in which an international programme had matured to deal with climate change and the ozone layer through the Montreal Protocol, and urged that this should be matched by international determination to work with developing countries on the problems of the tropical forest. There had been many international meetings in which tropical forests had been discussed, such as those of the European Commission and the G7. As a result, a number of priorities had been identified. The Commonwealth Prime Ministers' Conference had met in Malaysia, taking for its theme sustainable development as applied to tropical forests, and on which it had issued the Langkawi Declaration. There had been a series of detailed parliamentary hearings in Germany on the subject of tropical forests. The build-up towards UNCED was beginning.

Dr Freezailah also drew attention to growing opposition to the idea of bans and wished the Council to issue an authoritative statement on this issue to UNCED. ITTO's main issues were crystallizing: how could the sustainable management of tropical forests be most effectively brought about; what did sustainable management actually mean; how did it sit within the social and economic priorities of the tropical forest countries; what influence could the timber trade have upon the way in which the forest was managed; what were the relative responsibilities of the developed and developing countries in this regard; and how should ITTO priorities be determined? A searchlight was also being turned on the sustainable management of forests outside the tropics.

An important new issue appeared for the first time at this session. The UK delegation, in collaboration with FoE, introduced a project proposal for a labelling system for the promotion of sustainably produced tropical timber. This was the subject of intense and protracted debate, and much lobbying in the corridors. The proponents of the proposal affirmed that it was designed to help to secure a common basis of understanding on the actual requirements of sustainable forest management on the ground; it was seen as one practical step towards bringing this about. Perhaps more important, they looked upon it as a constructive proposal to draw the teeth of those who were still advocating a ban.[8]

The producers were privately alarmed that it would be difficult for them to provide such a label for *any* timber that they marketed. Publicly, they voiced many reservations and objections of variable force and validity. The proposals, as formulated, would not make a significant contribution to sustainable forest management given the more important and higher priorities for action in other areas. There would be great difficulties in implementing a system designed to define and evolve acceptable criteria for sustainable forest management and for standards for labelling timbers. The scope and thrust of the pre-project was too ambitious and did not take into account the interests of all parties in the tropical timber situation and economy. Its ultimate output was a veiled attempt to install a system which was an incentive to encourage the present campaign of boycotts against the import of tropical timber products, unless they could be proved to have come from sustainably managed tropical forests and were labelled as such.

In the end, the main objective of the proposal had to be abandoned. Instead the project was revised into one concerned with 'incentives in producer and consumer countries to promote sustainable development of tropical forests'. This was perhaps the first of several occasions on which the Council failed to show the necessary resolution and courage to provide leadership to the international community on an issue of great importance. It could, at this stage, have placed itself in the forefront of the debate about the interaction between environmental and social matters and free trade. In the view of many, it had failed to do so.

The NGOs became strongly critical. They were disappointed by the vacillation of the Council. The growth of ITTO was, in general, not in keeping with the imperatives of conservation. In particular, they singled out the 'compromise which had reduced the pre-project proposal of the UK to an insignificant study'. In addition, ITTO had 'so far failed to address the social problems'. The NGOs proposed an alternative programme of action.

The 8th Session of Council – Bali, May 1990

The 8th Session of the ITTC, held in Bali, was particularly important, not only because of its timing, but because it was being held in the country of one of the most significant (and, in the eyes of some, notorious) timber exporting countries. In the latter part of 1989 and early 1990, the campaign against tropical timber was spreading among some consumers; sanctions were being suggested as a way of stemming tropical deforestation; and there were preparations in hand for both UNCED and the 10th World Forestry Congress. In ITTO itself, the Sarawak Mission had taken place and the guidelines for the sustainable management of natural tropical forests were finalized. The time was ripe for ITTO to show its leadership. Bali saw some very significant developments: the first discussion of the report of the Sarawak Mission, which took place between November 1989 and March 1990; further moves to produce an integrated action plan; the birth of the Year 2000 Objective; and the first mention of the Bali Partnership Fund.

Guidelines

An important element of ITTO's work during and after the Bali meeting was the development of guidelines to provide norms for sustainable forest management. There are now (2002) five sets: *ITTO Guidelines for the Sustainable*

Management of Natural Tropical Forests (December 1990); *ITTO Guidelines for the Establishment and Sustainable Management of Planted Tropical Forests* (January 1993); *ITTO Guidelines for the Conservation of Biological Diversity in Tropical Production Forests* (September 1993); *ITTO Guidelines on Fire Management in Tropical Forests* (September 1997); and *Guidelines for the Restoration, Management and Rehabilitation of Degraded and Secondary Tropical Forests* (2002). These, taken as a set, are intended to provide a framework for national policies for sustainable forest management. In fact, as we shall see, they are, except the last, weak on aspects of land use in relation to forestry, on managerial functions, on the role of the private sector, and on consultation and collaboration with local communities. Some of these aspects were to be more thoroughly covered by the criteria and indicators.

The *Guidelines for the Sustainable Management of Natural Tropical Forests*

The *Guidelines for the Sustainable Management of Natural Tropical Forests* broke new ground. They laid down a number of principles and, within each, a number of possible actions. Like most policy documents prepared by ITTO, the guidelines were adopted by the Council from a draft prepared by an Expert Panel on the basis of a report by consultants. The accepted procedure was to choose two consultants, one from a producing and one from a consuming country, to obtain the benefit of different backgrounds and experiences (and to maintain the balance between the two groups). The core of the Expert Panel was also composed of equal numbers of members from producing and consuming countries supplemented by independent members with relevant experience, often from NGOs. There were, however, disadvantages in this procedure. Country experts often tended to consider themselves as representatives as well as experts. This could result in compromises which detracted from the force and clarity (and sometimes even from the logic, comprehensibility and good sense) of the resulting draft.

The decision to provide a shopping list of 'possible actions' rather than definitive guidelines was the result of such a compromise. The guidelines thus avoided the criticism of being too prescriptive – but were thereby weakened. Nevertheless, they were a great step forward; the 'possible actions' provided many valuable prescriptions and a checklist for the criteria and indicators which were developed later. The main headings are shown in Box 6.2.

An appendix on categories of forest land established the ITTO definition of PFE to include: production forest; protection forest on fragile lands; and forests set aside for plant and animal species and ecosystem preservation. It excludes forest destined for conversion to other uses.

Countries were invited to use the ITTO guidelines as a model upon which they might base national guidelines specially adapted to local circumstances. In fact, this seems to have interpolated an unnecessary phase and led to unnecessary delays in the all-important application of the guidelines in the forest.

BOX 6.2 GUIDELINES: PRINCIPLES AND POSSIBLE ACTIONS

POLICY AND LEGISLATION
Forest policy
National forest inventory
Permanent forest estate
Forest ownership
National forest service

FOREST MANAGEMENT
Planning
  Static and dynamic forest inventory
  Setting of management objectives
  Choice of silvicultural concept
  Yield regulation, Annual allowable cut
  (AAC)
  Management inventory and mapping
  Preparation of working plans
  Environmental impact assessment
Harvesting
  Pre-harvest prescription
  Roads

Extraction
Post-harvest stand management
Protection
  Control of access
  Fire
  Chemicals
Legal arrangements
  Concession agreements
  Logging permits on private or
    customarily held land
  Salvage permits
Monitoring and research
  Yield control and silviculture
  Environmental impact studies

SOCIO-ECONOMIC AND FINANCIAL
ASPECTS
Relations with local populations
Economics, incentives, taxation

Source: ITTO, 1990a.

The Year 2000 Objective

'By far the most important event at the Bali meeting, if not in the life of ITTO (to date, at least), was the adoption of the Year 2000 Objective. For the trade this was a major positive step in re-establishing the international reputation for tropical timber and tropical forests. It represented commitment to progress, as well as a very definite objective to tell the world about. It gave the trade renewed confidence in ITTO as an organization. There was a practical awareness that, even looking ten years ahead, there was unlikely to be a 100 percent success in achieving the aim. Nevertheless, it was a very cheering moment.'[9]

The first formal mention of the Year 2000 Objective, as it came to be known, was in the Draft Action Plan and Work Programme of the Permanent Committee on Forest Industry. The leading paragraph on Strategy reads as follows:

> Recognizing efforts of producing countries to sustainably manage and utilize their tropical forests and in order to contribute to the achievement of sustainable development in all producing countries, the objective is that the total exports of tropical timber products should come from sustainably managed resources by the Year 2000. This target date should be reviewed in 1995 in the light of progress on the

implementation of the Action Plan and Work Programme. Therefore, the long term development of appropriate forest based industries in producing countries is the central focus of the Strategy.

This is followed by a statement of the elements of the Strategy, which, in the context of that Committee, are all concerned with improvements in forest industry.

The producers, represented by Brazil, supported the target in these terms:[10]

> The tropical timber producing countries fully support the set target of the year 2000 as established in the Action Plan and Work Programme in the field of Forest Industry as well as the thorough implementation of the Plan.

The consumers, represented by the UK, went on record as follows:

> The consumers support the consensus on the adoption of the Action Plan on Forest Industry, and specifically the target date of the year 2000. We appreciate the magnitude of the task associated with achieving such an objective. However, we believe, in view inter alia of the support of producing countries for such a target, it will serve a useful purpose in supporting our efforts to promote sustainable forest management practices.

But a salutary note of caution was voiced by the representative of the US:

> In general, we question the advisability of adopting specific targets without a thorough understanding of the technical feasibility and economic implications of meeting those targets. With regard to the specific objective contained in the Action Plan that total exports of tropical timber should come from 'sustainably managed resources by the Year 2000', I note that the spokesperson for the producing countries has expressed her group's support for such an objective. In view of the fact that producing countries are in the best position to evaluate the target's feasibility, and must bear the ultimate responsibility for achieving such an objective with respect to the management of their tropical forest resources, we do not feel it is incumbent upon us to block consensus on the adoption of this objective. Nevertheless, we reiterate the view that, in general, such targets are not desirable. Further, we wish to add that, from our perspective, the establishment of such a target does not carry any implications for our government's trade policy.

In the course of the next few years, this 'target' subtly and significantly changed its emphasis (see, for example, Decision 3(X) given in Box 8.1 – the first Council-level decision on the Year 2000 Objective – or 'Target', as it was then called). These changes will be discussed later.

7

The case of Sarawak

'I am not worried about what happens to me, but what happens to my parents and grandparents'
A member of an informal deputation of the
Penan in evidence to the Mission

This chapter gives an account of the ITTO Mission to Sarawak – an example of how an ITTO mission works and what it can accomplish. It goes into considerable detail in order to illustrate how complex are the interrelations even in a relatively small state such as Sarawak. It reveals how difficult it may be to reconcile socio-economic development and the conservation of the resources upon which it must be based, but it also demonstrates how it proved possible to disentangle some of the complications and select a few of the most salient points for action.

This account is based on the report of the Mission (ITTC(VIII)/7) and on personal recollections. It uses quotations from the full report and extensive excerpts from a shortened version prepared by Lord Cranbrook. It is an account of the situation as it was in 1989–1990. The chapter concludes with reactions to the report within the ITTC and with an account of the action that ITTO has since been able to take to assist Sarawak. There is, unfortunately, no independent up-to-date account of the present situation that might be used to judge the Mission's long-term effects.

The Mission

The composition of the Mission, and especially the choice of its leader, was an important matter for the Council. Members of the Mission had to be acceptable to the Federal Government of Malaysia and the State Government of Sarawak, but they also had to satisfy producers, consumers and other interest groups within the Council. The Earl of Cranbrook was chosen as Leader – a distinguished zoologist who had worked in the past in the Sarawak Museum and the University of Malaya in Kuala Lumpur, who spoke Malay and knew Sarawak well – and who was known and respected there. The nine other members were chosen to reflect the membership and interest groups in ITTO. Three were from producing members, three from consuming members, one from the timber trade, one representing the NGOs, and an independent economist. The last two served as joint rapporteurs. The Executive Director of ITTO accompanied the Mission throughout. The terms of reference were set by Resolution 1(VI) reproduced in Box 6.1.

This Mission took place between November 1989 and March 1990. Members paid two short visits to Kuala Lumpur to meet the Prime Minister and other federal ministers, and to visit the Forest Research Institute Malaysia (FRIM). They spent some six weeks in Sarawak, spread over three visits. The programme of the first visit was organized in Peninsular Malaysia by the Ministry of Primary Industries and in Sarawak by the State Ministry of Resource Planning. The programmes for the second and third visits to Sarawak were decided by the Mission itself.

The Mission worked through field visits, meetings, dialogues and the inspection of documents and literature. Among other places, visits were made to relevant government departments, mixed hill dipterocarp forests, swamp forests and mangrove forests, timber concessions, logging operations, research plots, forest industries, ports and industrial development areas, training institutions, agroforestry projects, national parks and wildlife sanctuaries, Divisional and District offices, rural villages and longhouses. The Mission was assured of, and in the main had, complete freedom to go wherever it desired and to meet whoever it wished. Records of all the visits and discussions were appended to their report.

Meetings were held with the timber trade through the Sarawak Timber Industry Development Corporation (STIDC) and the Sarawak Timber Association (STA). The views of local organizations, representative community leaders and other members of the public were obtained through meetings and dialogues held in Kuching, Lundu, Sibu, Kapit, Miri, Limbang, Marudi and Long Teru, Tinjar. The locations of public dialogues were advertised in advance. Special river transport was provided to enable the attendance of people from remote areas. During parts of the second and third visits, the Mission divided into groups in order to cover the ground as thoroughly as possible. These concentrated respectively on forest management, forest industries, and the environmental and social aspects of sustainable forestry. Members were therefore able to gain an exceptional insight into forestry in the State and the views of the many groups of Sarawak society about the way in which it was conducted. The programme was organized by the State Government with great efficiency; the openness and candour of the hosts were outstanding.

The Task of the Mission

The first section of the report of the Mission set out some of the background to the study: the controversy over the use and conservation of tropical forest; the criticism of the policies of Malaysia and Sarawak, the competence of the Mission and the way in which it should interpret its terms of reference; the 'native peoples' question; the meaning of sustainability; and a summary of how the Mission saw the task facing it.

As these questions were fundamental to the way the Mission approached its work, they were extensively discussed at an early stage. It should not be assumed that there was always complete agreement between the members of the Mission; indeed, the cross-currents of opinion were vigorous and stimulating. In the early stages, some members were so impressed with the professionalism of the Sarawak Forest Department (perhaps in comparison with their home organizations) that they felt that the Mission should not criticize. But members later accepted the argument that their hosts wanted them to criticize, provided that the criticisms were constructive; otherwise, why invite the Mission? The Sarawak authorities hoped that an objective assessment of the quality of their management might help to stem the flood of international criticism, which they felt was unjustified. Other Mission members were suspicious that the leader and one of the rapporteurs, both of whom had environmental credentials, would unduly influence the final conclusions. But, in the end, the Mission itself became an exercise in international understanding. Indeed, Señor Muñoz-Reyes of Bolivia, who was initially one of the most sceptical, has since claimed that he was 'converted' by the Sarawak Mission; he went on to persuade the Government of Bolivia to request a similar ITTO mission there. At the meetings in Sarawak, a working rapport developed between the various interest groups. The conclusions and the reasons for them were unanimously accepted, although one highly critical issue – the recommended annual harvesting rate – represented a negotiated compromise between strongly held views.[1]

As its main work was on the 'sustainability' of the forest resource of Sarawak, the Mission developed a working definition of sustainability, as follows:

> 'Sustainability' is taken to mean the attainment and maintenance of a forest estate yielding a continuing and non-declining flow of benefits and products at levels of each considered best by the people of Sarawak and at levels which can be sustained or even increased in the future. When policies and practices for forest management are formulated, full account has therefore to be taken of the potential for increasing the yield of each individual benefit or product and this potential should be preserved as far as possible.

Accordingly, the Mission decided that it should examine separately five aspects of sustainability which, in the view of the members, added up to a comprehensive interpretation of sustainability within their terms of reference. These separate aspects were: (1) the capacity of the forests of Sarawak to provide a sustained supply of timber in the medium and long term, both for domestic use and for export; (2) forest management for sustainable timber supply; (3) the maintenance of soil and water quality; (4) the maintenance of biological diversity in the forests of Sarawak; and (5) the economic sustainability of forestry and the forest industry.

The 'Native Peoples' Question

Although they do not figure in the list above, social aspects were not neglected. Members were very exercised about how to handle them, for these were clearly matters of great delicacy and there was a danger that the whole report might be rejected if they were treated insensitively. That must be avoided at all costs. On the other hand, where the Mission had concerns, these should be voiced. One alternative was to include a section in the report that dealt specifically with social issues; but, after much debate, this was rejected in favour of including social issues within each of the other sections. This, in turn, affected the way in which the final recommendations were formulated.

As the question of the traditional rights of the indigenous (ie non-Chinese) rural communities (in Sarawak terms, denoted as 'natives') to and in forested land has figured so largely in criticisms both of Sarawak policy and of the Mission's report, this section of the report is quoted at length:

> To some of the critics this is the vital issue in the whole controversy – 'the whole reason that the timber industry in Sarawak has become so controversial is because the native people have been actively resisting the logging of what they perceive as their traditional lands'. Other critics, while not going as far in anchoring their concern solely to traditional rights, do see them as an essential element in their opposition on conservation grounds to current policies and practices. In such circumstances, not to address this issue in a serious way would undermine the credibility of the Mission as an independent and objective body and any belief in ITTO as an instrument for the betterment of the world's tropical forests.
>
> It is, in fact, the only point in the dispute on which one or other of the opposed views alone can be right. Either the native communities have rights which are being or could be violated by logging or they do not. Yet, strictly speaking, it is not, in itself, a matter for the Mission to comment, let alone pass judgement, on. More significantly, neither is it a matter for the ITTC under whose authority the Mission is established. There is no way in which the International Tropical Timber Agreement can be interpreted as extending into such areas. In other words, it can be argued, with some justification, that this is none of the Mission's business. The rights and wrongs of the traditional rights issue are matters for determination in the courts of Malaysia. The Mission is neither qualified nor empowered to judge on such matters.
>
> Nevertheless the Mission does not unreservedly accept that the question of the traditional rights is none of its business. In the light of the very clear responsibility which the International Tropical Timber Agreement places on ITTO in the conservation and management of the tropical forests and the Mission's task to assess 'the sustainable utilization of tropical forests ... in Sarawak', there is a sense in which it could be very much its business... If the extent, nature and location of these rights, and how, when and where they are exercised, affect either

or both the area and productivity of the forests, then native community rights are relevant to any assessment of sustainable levels and the quality of performance. That aspect cannot be avoided, ignored or passed off as the responsibility of somebody else. If it arises in this form, the Mission must deal with it.

The simple expedient would be to withdraw the areas in question from timber harvesting operations and planning, at least until the legal position is sorted out definitively. There would be a reduction in the long-term timber flow and that would, in turn, necessitate recalculation, relocation and reallocation of concession areas. Some difficulties with logistics, contract renegotiation and compensation would, no doubt, arise but they could hardly be taken as being insuperable. They may, of course, take some time to overcome and that would necessitate some transitional arrangements. But the claimants and their supporters would get what they claim is theirs, while the cost in long-term sustainable timber output would not be all that great.

Although the information available to the Mission is somewhat approximate it does not appear that such reservations would total more than about 0.5 million hectares. The implicit reduction of around 15 per cent in the long-term timber flow is probably well within the statistical confidence limits of the estimate of that flow and could not, by any standards, be regarded as catastrophic.

This simple expedient would therefore be unlikely to have serious consequences for sustainable timber utilization, other than perhaps over a relatively short transitional period. On the other hand, it would greatly enhance the prospects for sustainability of certain social values and for those non-timber ecological values which are not at risk under shifting cultivation. At the same time it would remove the issue of traditional rights as a plausible rallying cry in the debate. On those grounds it can hardly be faulted, except for the implication that the government would be seen to be compromising its authority and sovereignty under duress. That, however, does not necessarily follow at all… Both the constitution and the forest policy guarantee traditional rights and the status of those rights is presently under review in the courts.

If this simple expedient were to be adopted, there would obviously be some reduction in the sustainable yield of timber. Whether the reduction were temporary or permanent, the Mission would have, at least, to consider its effect on 'sustainable utilization'. The question of the native community rights themselves would not, however, arise. Even if the simple expedient is not adopted, then a continuation of the present situation, at least for some time, will still have to be taken into account. This does not mean that the Mission would therefore have to take sides in the issue or judge on it. But, for the purpose of making proposals for the improvement of the quality of sustainable management, it would certainly need to consider possible remedies. And that may take it into an examination of causes.

Forestry in Sarawak

The report then goes on to describe the constitutional position of forestry in Sarawak, the Sarawak Land Code, the Sarawak Forest Policy, forest legislation, forest management plans, research and training, and the organization and the development of Sarawak forest industry at that time (ie 1988–1989). Some aspects of these are important for understanding the nature of the Mission's recommendations.

The Constitutional position

In the Constitution of Malaysia, the legislative and executive authority over forest is a State responsibility. The role of the Federal Government[2] is mainly confined to research and development, maintenance of experimental and demonstration centres, education and training, forest industries development, and technical assistance to States in terms of overall forestry development and management.

The Sarawak Land Code

Dispute over native customary claims to land rights had been a major cause of the internal and international tensions that related to logging in Sarawak. It was, therefore, essential to understand clearly the background and present operation of the relevant laws and regulations. There had been continuity in the law of Sarawak through the Brooke regime and the subsequent period of British administration to modern Malaysia.[3]

Historically, the traditional method of subsistence agriculture followed by the majority of the settled inhabitants of the interior of Sarawak had been shifting cultivation for an annual crop of hill rice. The chief way in which customary land rights were acquired was by the felling of virgin jungle and the subsequent occupation of the land. Recognized rights might also be obtained by other actions such as the planting of fruit trees and use for funereal purposes. These rights were backed by native customary law, which differed only in minor detail from one community to another.

The report drew attention to the history of Sarawak law relating to land, starting from the earliest legislation of the first Rajah Brooke. The Land Regulations, 1863 gave the Brooke Government ownership of 'all unoccupied and waste lands' (which could thenceforth be leased or sold) but was not intended to affect land already occupied by any native of Sarawak. However, the need for amendments in 1871 and 1875 that gave certain rights to 'squatters' suggested that even the comparatively simple 1863 regulations did not fully accord with the existing practices of customary tenure.

The first comprehensive legislation was introduced in 1920 (Land Order no VIII), but this was later repealed and replaced by the 1931 Land Order, which recognized Native Areas – to be inhabited only by the indigenous races of Sarawak – and Mixed Zones, in which the Chinese and others could hold rights. The subsequent Land Settlement Order (1933) proposed drawing boundaries around longhouses, within which members of the community in question would have exclusive rights to establish customary tenure, and to appoint village councils charged with the task of resolving disputes over customary tenure. This Order, in turn, was followed by a Secretariat Circular on land tenure, no 12/1939.[4] Appended to the 1939 Circular was the Senior Forest Officer's memorandum, illuminating the perceived conflict between shifting cultivation and forestry and proposing a tripartite land use classification, into Protective Forest, Productive Forest and Agricultural Land. Together, these documents reflect efforts at the time to regulate shifting cultivation and to resolve land disputes within and between communities, especially those arising from internal migrations.

After Cession (1946), the incoming British colonial administration consolidated the law, enacting in 1948 the Land (Classification) Ordinance which recognized five categories: Mixed Zone Land, Native Area Land, Native Customary Land, Reserved Land and Interior Area Land. Subsequent measures were introduced in 1954, whereby no rights could be created by occupying Mixed Zone or Native Area Land, and in April 1955, whereby new rights within Interior Land could only be created under a permit issued by the District Officer in whose district the land was situated. In 1957, further legislative steps were taken to consolidate and update the laws relating to land.

The legal framework, within which alternative, often competing, uses of land are now regulated in Sarawak, is therefore the Land Code Ordinance of 1958 (Sarawak Cap 81).[5] Although a challenge to some aspects of the law concerning native customary rights was before the courts at the time (1990), the Mission was told that this text could be taken to be a correct version.

The Land Code retained the five land classifications of its 1948 predecessor. Its provisions referred mainly to non-forested land. This could be either Titled or Non-Titled. Titled land posed no problems. The Non-Titled could be land held under Native Customary Rights (NCR) or Native Communal Reserve. The accepted means by which NCR might be acquired were set out in Section 5 (2), but the 1955 regulation (above) was formalized by Section 5 (1) as follows:

> As from the 1st day of January, 1958, native customary rights may be created in accordance with the native customary law of the community or communities concerned by any of the methods specified in subsection (2), if a permit is obtained under section 10, upon Interior Area Land. Save as aforesaid, but without prejudice to the provisions hereinafter contained in respect of Native Communal Reserves and rights of way, no recognition shall be given to any native customary rights over any land in Sarawak created after the 1st day of January, 1958, and if the land is State land any person in occupation thereof shall be deemed to be in unlawful occupation of State land.

Section 10 (4) specified that:

> The occupation of Interior Area Land by a native or native community without a permit in writing from a District Officer shall not, notwithstanding any law or custom to the contrary, confer any right or

---

BOX 7.1 THE LEGAL STATUS OF NATIVE CUSTOMARY RIGHTS LAND

FOREST
NON-FOREST
   Titled
   Non-titled
      •   Land with Native Customary Rights (NCR)
          –   'Held', if registered by 1958; or, if felled forest, between 1958 and 1972, registered under Section 10 of the Land Code
          –   If not felled, the land cannot be legally 'held'
      •   Native Communal Reserve – gazetted – can be converted into titled land

---

privilege on such native or native community and, in any case, such native or native community shall be deemed to be in unlawful occupation of State land.

Provisions were made for the registration of NCR land under rights obtained before 1958 or, in the case of felled forest only, if registered between 1958 and 1972 under section 10 of the Land Code. Until a document in title had been issued, such land continued to be State land, but any native lawfully in occupation of it was deemed to hold by licence and was not required to pay a rent. It was therefore illegal under the Land Code to occupy Interior Area land which had not been felled for cultivation (or otherwise legally occupied) before 1 January 1958. Rights to lands by natives might be obtained through Native Communal Reserve, which were gazetted and held under joint ownership by the village community. They could be converted into land with individual titles in perpetuity after detailed survey.

The position presented to the Mission, is summarized in Box 7.1.

The Sarawak Forest Policy

The Forest Policy of Sarawak in 1988 had originally been formulated and adopted in 1954 (a time when the British colonial administration was in power). The general statement of policy read as follows:

1 To reserve permanently for the benefit of the present and future inhabitants of the country, forest land sufficient:
(a) for the assurance of the sound climatic and physical condition of the country; the safeguarding of soil fertility and of supplies of water for domestic and industrial use, irrigation and general agricultural purposes; and the prevention of damage by flooding and erosion to rivers and to agricultural land;
(b) for the supply in perpetuity and at moderate prices of all forms of forest produce that can be economically produced within the country, and that are required by the people for agricultural, domestic and

---

Box 7.2 STATUTORY DESIGNATIONS OF FOREST LAND

PERMANENT FOREST ESTATE
    Forest Reserves
    Protected Forests
    Communal Forests
TOTALLY PROTECTED AREAS
    Wildlife Sanctuaries
    National Parks
FORESTS ON STATE LAND

---

industrial purposes under a fully developed national economy.

2 To manage the productive forests of the Permanent Forest Estate with the object of obtaining the highest possible revenue compatible with the principle of sustained yield and with the primary objects set above.

3 To promote, as far as may be practicable, the thorough and economical utilization of forest products on land not included in the Permanent Forest Estate, prior to the alienation of such land.

4 To foster, as far as may be compatible with the prior claims of local demands, a profitable export trade in forest produce.

An understanding of the current statutory designations of forest land is also important for understanding the recommendations of the Mission. The various categories of forest are outlined in Box 7.2:[6]

The PFE and the totally protected areas (TPA) are under the jurisdiction of the Forest Department. The remaining forests are those on State land over which the Forest Department has no legal jurisdiction after their timber is removed. All forests, however, are subject to licence to extract and all have management plans.

Findings of the Mission

Sustainability of timber yield

It was most important that the Mission should determine what yield was likely to be sustainable from the forests of Sarawak. This depended upon (a) the area available in perpetuity within the State for timber production (the PFE); and (b) the expected yield from that area. The report goes into detail on this crucial aspect. *The final figures calculated for sustainable timber production were only to be considered valid if the conditions attached to each were observed.*

Available land

The Mission calculated the area of land that might be available for timber production under various assumptions. There might be competition with

Table 7.1 *Land available for sustainable timber production*

Forested lands available (including peat swamp and mangrove)	Area in production (ha)
PFE <60% slope	3,135,000
Add State <60% slope	4,513,000
Add PFE >60% slope	5,198,000
Add State >60% slope	5,783,000

production forest for permanent use of the land – competition from catchment protection, the protection of fragile soils, the preservation of biological diversity, traditional shifting cultivation and conversion to plantation agriculture. Decisions about how much land should be allocated to each of these uses would be fundamental. The Mission estimated[7] that foreseeable requirements of land for purposes other than timber production[8] would exclude more than half of the land area of the country, leaving approximately 5.9 million hectares out of a total land area of 12.4 million hectares. Of this 5.9 million hectares, about 1.4 million hectares was still State land and might be destined for further agricultural development. Moreover, at least one-sixth of the PFE was on slopes steeper than 60 per cent, where present logging methods were incompatible with the catchment objectives of the Sarawak Forest Policy.

In some parts, notably southwest Sarawak, there was already a land shortage for traditional agriculture, making it difficult to establish or retain land as PFE or even as TPAs. This pressure was being relieved to some extent by agroforestry schemes and by agricultural development schemes. But daily paid labour was regarded as a poor alternative to the independence of traditional agriculture. Meanwhile encroachment continued, mainly into logged-over forests, at rates of around 5000ha per annum for PFE and about ten times as much on State land. Shifting cultivation was probably diminishing in the long run but had to be taken into account.

From these figures, it was tentatively estimated that the land that would be available for sustainable timber production in the future, under the four different assumptions, would be as shown in Table 7.1.

Potential yields

The other important characteristic was potential yield. To assess this, possible lengths of cutting cycle (for the hill dipterocarp forests) were calculated, based on size at maturity, species and treatment. If the harvest were limited to trees of 60cm or more diameter at breast height (dbh) and species of the highest wood quality, long cutting cycles would be necessary to attain the yield at that time of 38 m^3/ha per annum. A minimum dbh of 45cm, taking species of lower quality, would allow a 35-year cutting cycle. The latter figure was used in the yield predictions. Its validity depended on the assumption (although this was of course uncertain) that future markets could be found for smaller logs and more diverse species. Furthermore, this yield could be increased and the length of the cutting cycle reduced by 'liberation thinning' (see below).

Additions were made for the potentially productive peat-swamp and mangrove forests: for swamp, 1,370,000 m³ per annum from PFE and 870,000 m³ per annum from State forests; for mangroves, 111,000 m³ per annum from PFE and 258,000 m³ per annum from State forests.[9]

Possible sustainable yields

The conclusion, therefore, was this. If allowance were made for other anticipated land uses and if the present policy of excluding lands of over 60 per cent slope was maintained, it would only be possible to sustain a yield of about 4,100,000m³ per annum. The Mission judged that it would be unrealistic to recommend a precipitate reduction in the rate of harvest to this level because of its social and economic impacts on Sarawak. Accordingly, they explored other options for sustaining a higher yield. There were two possibilities: increasing the available area or enhancing the yield.

By increasing the available area

Apparently 824,000ha of State land were not needed for other uses. If this were to be converted to PFE and used for continuing timber production, the sustainable yield would be raised to about 6,300,000 m³ per annum. But this would entail the gazetting of large areas of State lands as PFE – a process that had proved difficult and time consuming.

Within the existing PFE there was a logging restriction on slopes steeper than 60 per cent. If this restriction were to be lifted, the available area would rise by 683,000ha and the sustainable yield would reach 7,000,000 m³ per annum. But, with current logging practices, this gain would produce unacceptable environmental damage. If such land were to be added to the productive timber land base, reformed practices would be needed, possibly involving new equipment and higher costs.

An additional area of State lands steeper than 60 per cent slope was apparently available. If this 588,000ha were added to the productive area as PFE, the sustainable yield could be raised to about 7,700,000 m³ per annum. This would depend not only on the successful gazetting of these lands but also on the assurance of a much higher quality of logging practice than was traditional.

By increasing the yield

Liberation thinning had already been applied experimentally to some 35,000ha and had significantly accelerated the growth of selected trees in cut-over forests. The treatment was costly and required trained crews, but it promised not only to provide much-needed added sustainable yield, but might also reduce the length of the cutting cycle by as much as ten years. But liberation, if it were to be effective, would have to be applied promptly in all logging areas, and would call for a silvicultural effort so far made only on an experimental scale.

The possible sustainable annual yields derived from these calculations were as shown in Table 7.2.

Table 7.2 *Potential sustainable annual yields*

Forested lands available (including peat swamp and mangrove)	Sustainable annual yield (m³)	
	Untreated	Liberated
PFE <60% slope	4,100,000	6,300,000
Add State <60% slope	6,300,000	9,200,000
Add PFE >60% slope	7,000,000	10,500,000
Add State >60% slope	7,700,000	11,000,000

The consequences of 'business as usual'

If harvesting of the hill forests continued at the rate at the time of the Mission of about 13 million m³ per annum, all the primary forests in PFE and State land which were assumed to be available for timber production, including those steeper than 60 per cent, would have been harvested in about 11 years. Only cut-over forests would remain. After that, there would be a sharp decline in yield, employment and revenue until the cut-over forests matured again. This serious prospect might be alleviated if a start were made immediately to reduce the rate of cutting so that the remaining primary forests could provide timber until the cut-over forests matured.

The effect of present timber extraction on future yield

Furthermore, these predicted sustainable yields could never be attained by continuing many prevailing practices. Logging damage to residual trees was excessive.[10] Marking of stems to be left and more careful felling (with penalties for damaging residuals) would do much to improve the situation. The practices of the labour force in the forest caused much of the damage. There was practically no formal training for fellers or tractor and skidder drivers; experience was passed from one to the other. Tractor and skidder drivers were paid piece rates by most companies, and fellers apparently by all companies. Emphasis was on output. It was hardly surprising, therefore, that little attention was paid by the fellers to limiting damage to the residual stems or by skidders to this or other effects on the environment. Safety procedures were also usually of a low standard. These weaknesses were exacerbated by inadequate staffing of the Forest Department and the consequent inability to exercise the degree of supervision needed.

Logging and local communities

Shifting cultivation had continued to reduce the area of forest from which future timber yields could be expected. According to statistics provided by the Forest Department, 116,000ha within the PFE had been lost to shifting cultivation by 1985, and 1,080,000ha of primary forest in Interior land (State land) had been cleared for shifting cultivation between the 1960s and 1985.[11] Clearance was still proceeding, before and after logging; and no instance of Section 10 procedure was produced, although members of the Mission asked to see examples.

The cost of logging was also increased by disputes over the rights and interests of local communities. This endangered both the economics of current operations and long-term economic sustainability. In the dialogues and informal discussions held with leaders and other representatives of local communities, no objections, with a few exceptions, were raised to logging as such, but rather to its speed and where and how it was conducted. Many recognized its importance to the economy – the advantage of jobs in logging, and the fact that some of the revenue funded social projects. There were strong complaints of invasion of 'temuda' (former farmland claimed under NCR), violation of cemeteries, damage to domestic water supplies, to watercourses and consequently to fish stocks, depletion of game, the disturbance of paths through the forest, loss of timber trees needed for personal use (eg boat building), destruction of rattan which was valued for domestic use and as a traded commodity, felling of fruit trees and of protected tree species of commercial value, such as 'engkabang' and – in the case of Penan – of the poison trees ('ipoh'). Grievances included lack of consultation and unsympathetic treatment by operators and contractors, failure to be offered work except at menial levels, social problems arising from the intrusion of outsiders, and little direct participation in financial returns. There was an underlying feeling that the forest belonged to the local people and that the costs fell on them while the benefits went elsewhere. It was claimed that the blockades arose from the failure to obtain a hearing through the proper channels.

In their representations, local community leaders and others made no distinction between native customary rights recognized under the Sarawak Land Code and claims that would not be admissible under Section 5. The requirements preceding gazetting of Forest Reserves, Protected Forests, National Parks and Wildlife Sanctuaries involved detailed evaluation of all claims to customary rights, undertaken by the administration in conjunction with traditional community procedures. Existing rights might be recognized in the gazettment and allowed to continue in an appropriate manner or be disallowed, with or without compensation. Hunting rights might also be admitted in forest land that had not been cleared or otherwise claimed under NCR.

Because all claims were settled in this manner before gazetting of PFE or TPAs, any subsequent clearance of the land for farming or other purposes could not, under Sarawak law, give rise to new claims under NCR. Hunting within TPAs or Forest Reserves was illegal unless rights had been recognized; but extensive rights to hunt and gather forest produce were available for all Sarawak Natives within Protected Forests.

The Mission therefore concluded that many of the present complaints and claims of right by local communities were not based on Sarawak law as established by the Land Code but on a continued application of older, traditional native custom. It was not apparent that this distinction was appreciated by the communities concerned or, indeed, by others pressing the case.

Compensation always turned out to be an important item on the agenda. To meet the situation, in many instances 'sagu hati' (goodwill) agreements had been offered by logging companies to local leaders or to entire communities, specifying (often in great detail) arrangements for compensatory cash payments. Although strictly illegal, these were becoming formalized.

There was also considerable interest in the establishment of more Communal Forests. There were administrative objections to using this means to enhance community participation in logging – responsibility passed out of the hands of the Forest Department. But the rights of local communities to forest products for their own use could be met, for example, by reserving parcels of Protected Forest near longhouses for local use. The Forest Department could still maintain control and ensure sustainable utilization.

Environmental sustainability

Catchment management

On catchment management, the report singled out the severe gully erosion in many concessions and the resulting siltation of rivers. Recommendations included: the halting of extraction on steep and erodible terrain; the need to make legislative provision for protection forest; a code of environmental conduct covering standards for the design and construction of roads; stricter application of the 30 per cent slope rule; the establishment of riverine reserves; the treatment of abandoned roads and logging tracks; and the formal training of operators, to be based on a model concession used for training, demonstration, research and the development of new logging methods.

Biological diversity

The section on biological diversity reviewed the particular biological richness of Sarawak and the legal provision for its protection. There were already plans to double the TPA network to over 1 million hectares (8.33 per cent of Sarawak's land area). Even with these proposals, it was considered that the TPA system would be inadequate. The report recommended a number of additional areas – two national parks where the majority of nomadic Penans lived, areas for the effective protection of orang utans and white-fronted langurs, areas to preserve a sample of the inland flora of the zoogeographical region south of the Rajang River, areas to protect the highly endangered Sumatran rhinoceros, and the preservation of a complete peat swamp system. Additional recommendations covered measures for improving the land use setting of TPAs, the management and protection of TPAs (for example, by more rigorous prevention of illegal incursion linked with better opportunities for productive farming outside and better public relations), better information about the advantages of TPAs, strengthened staffing and resolute leadership.

Economic sustainability

There was also a long, closely argued discussion on the economic sustainability of Sarawak's forestry and forest industries, taking the term to mean 'that the economic structure built upon and around the utilization and management of the forests must be able to continue indefinitely at not less than the present level'. It reviewed the chain of forest industries involved in the growing, harvesting, transporting, processing and marketing of wood (and possibly bark);

the network of industries concerned with the utilization and further processing of the output in Sarawak, and with supplying materials and services to those industries; the infrastructure developed wholly or partly in support of the growth of those industries; and the State revenue base. Also considered were the non-wood products and services of the forest.

The conclusion of the Mission was:

> that present policies, strategies and practices in respect of timber production [were] not economically sustainable. This was certainly so from the forest resource side and possibly also from the market side. The Mission [was] not able to make a reasoned assessment of the economic sustainability in terms of non-wood products and services.

There were doubts about:

> the economic sustainability, under present practices, of water quality and river transport systems without massive and increasing investment. Any adverse effects on fish populations could be long lasting, but for all non-wood products the evidence [was] conflicting and indefinite. On the whole, the Mission was inclined to accept as highly likely that policies as presently implemented were reducing the capacity of the forest resource to sustain the rural economies based on them, unless protective measures were put in place.

The Assessment

The Mission concluded that forest management in Sarawak was undoubtedly of a much higher standard than in most other tropical timber producing countries and even in some developed countries. Yet, the present utilization of the forests of Sarawak was not fully sustainable in certain important respects. There were great differences, even on land that would be retained as forest. For example, management in the peat swamp forests was effective for all species except 'ramin' (*Gonystylus bancanus*); for the hill dipterocarp forests, management planning was good but execution poor.

Exemplary features were: a policy that set watershed management as one of the primary objects; the reservation of totally protected areas for wildlife sanctuaries and national parks; an outstanding system for tracing and controlling the movement of logs from the forests to mills or export points; the simple but comprehensive management planning for the production forest in the PFE; and the comprehensive research data bank of vital management information. The tracing system was 'outstanding by any standards', and the data bank was 'almost unequalled in extent and quality'.

Possible improvements would be the updating of policy to reflect constitutional changes over the last 36 years and recent changes in forest values, and to include national parks and wildlife protection in the formal statement of forest policy. The TPA network was too small and inadequately controlled.

Growth rate and yield estimates were monitored well but the information was not used. Management prescriptions, particularly for logging operations, were insufficiently specific and rigorous.

Implementation was weak. Sustainable management was generally successful only in terms of conservation, biological diversity and ecological balance in the National Parks and Wildlife Sanctuaries, and in timber production in the peat swamp forests. Even in these there were significant deficiencies. Effective safeguarding of the fully protected areas depended upon the human communities in the vicinity being able to survive without being forced (by economic pressures) to encroach on them. That depended, in turn, on the success of economic and social development in the State in providing alternatives. That, in turn again, largely depended on a successful and sustainable forest industry, which depended upon successful sustained yield forest management for timber production. This chain of dependencies would fail if sustained yield management for timber could not sustain the industry built up in anticipation of it. Pressure could subsequently build up to open the TPAs for 'limited' harvesting to save the established industry and its dependants from economic collapse.

The two most evident weaknesses were the inadequate standard of catchment management and the overcutting of hill dipterocarp forests, the output of which could not be sustained at anywhere near the current level. These forests were, and would continue to be, the main forest resource of the State; if sustainable management of these were to fail, almost every object of the stated policy for sustainable use of the State's forests would be at risk. The present failures were caused mainly by the under-staffing of the Forest Department and the arrangements through which timber licences were issued, managed and operated. For example, while the output from Hill Forests had risen over the past five years from about 9 million m^3 per annum to about 13 million m^3, there had been no increase in the staff of the Forest Department. This shortage affected all aspects of the department's responsibilities. Also, the scale of concession allocation was much higher than any prudent estimate of the sustainable productive capacity of the PFE.

Another damaging feature of the concession system was the structure of the hierarchy whereby the licensee – the concessionaire – sub-contracted his rights to a second party, who in turn sub-contracted the operation to a third party who employed the actual loggers on the basis of payment by output. There was also no synchronization between the length of the licence period and the forest management cycle. The practice of issuing licences for shorter than the felling cycles, without any guarantee of renewal, was not helpful to long-term management. It tended to reward those concessionaires and operators who planned their operations so as to get the maximum possible output with the minimum possible fixed investment, regardless of the effects on the viability of the concession over the remaining years of the felling cycle; and it penalized those who attempted to average out the cost of logging over the concession area as a whole. Short-term licences were also a strong disincentive to licensees to invest in timber processing.

Recommendations

Although it was preceded by a short summary written personally by the Mission leader, the report did not include an executive summary. From the wide sweep of matters treated in the main body of the report, including several which emphasized deficiencies and proposed remedies, the matters chosen for formal recommendation were presented under half a dozen heads. As a result, some observers considered the final conclusions and recommendations (summarized in Box 7.3) surprisingly limited. Three were directed towards the Governments of Malaysia and Sarawak, and three towards the international community.

Perhaps the most critical (and certainly the most criticized) recommendation was the figure for a State-wide 'sustainable' rate of harvest of 'about' 9.2 million m^3 per annum. The derivation of this figure requires a careful study of the report, where it is made clear that such a level of harvesting can only be justified if there is a concomitant commitment to effective silvicultural management of the PFE. The recommendation itself is not silent on this aspect but, unfortunately, parties in the controversy have habitually quoted the figure alone, without its attached constraints.

In retrospect, it was perhaps an error of judgement to simplify the final recommendations by not presenting a complete consolidated list. It might also have been more convincing to present the various possible rates of sustainable harvest with the qualifications attached to each. It is unwise to assume that any report will be read fully and conscientiously!

Reception at the International Tropical Timber Council

The Report was released at the Bali meeting (May 1990) and some preliminary comments were received; but full consideration was postponed until the next meeting in November 1990, when members and others would have had full time to consider its findings. Nevertheless, most of the environmental NGOs had already made up their minds to discount the report. For example, Fay of the Environmental Policy Institute observed:

> The NGO community had considered the ITTO Mission to Sarawak as an historic opportunity to observe the problems of Sarawak on the ground and make recommendations to promote sustainable forest management. They had, therefore, felt disappointed at the findings of the Mission which gave clear priority to timber production over all other considerations – and this contrary to the declared official policy of the Sarawak authorities which emphasized the prior claims of local demands over the export trade in timber. The ITTO should, therefore, carefully consider its position on the report.[12]

This, and the later treatment of the report by the NGOs, was an early example of their failure to build upon an ITTO achievement which, though not perfect

BOX 7.3 MISSION RECOMMENDATIONS

A *For action by the Governments of Sarawak and Malaysia*

The Mission recognizes that there are many admirable features in Sarawak forestry. It believes that the sustainable management of the forests of Sarawak is being partly achieved, but full achievement depends on immediate action in three aspects:

1   Firstly, the staff of the Forest Department must be comprehensively strengthened.
2   Secondly, the annual rate of harvesting must be phased down to a figure that corresponds to the prospective sustainable yield, ie for a Permanent Forest Estate of 4.5 million hectares of land at slopes of less than 60 per cent, of which a substantial portion is silviculturally treated, plus State land forest not allocated to other uses within the same slope limitation, about 9.2 million m^3/yr.
3   Thirdly, the standards of catchment protection in the hill dipterocarp timber production forests must be improved.

Recommendation A1 – Strengthening of the Forest Department
Increased staffing of the State Forest Department is essential if the economic potential of the forest resources of Sarawak, its forest industries and other forest values are to be maintained. Without strengthening, the commendable Forest Policy adopted by the State Government cannot succeed, nor can the other deficiencies identified by the Mission be properly remedied.

To effect these fundamental improvements, immediate strengthening is essential in two fields: control and planning.

(a) *In control.* To enable effective implementation of all aspects of harvesting, forest management and environmental protection, urgent attention needs to be directed to control on the ground: to ensure strict adherence to the conditions of concessions, including the prescriptions in management and engineering plans; to ensure high standards in road building and harvesting operations; to prevent illegal encroachment into the Permanent Forest Estate and into Totally Protected Areas; and to suppress illegal hunting.

(b) *In planning.* To develop a basic plan for the long-term sustainable management of the productive forest resources of Sarawak and the development of its forest industries, based on a reliable assessment of the forest resources in the State and of the long-range demand and supply outlook for forest products.

These measures should be supplemented by strengthening in research, education and training, and public relations.

(c) *In research. Silviculture:* to make full use of existing research data and to extend research into a more fully representative series of sites in the hill dipterocarp forest. *Harvesting:* to investigate alternative methods of harvesting that are more efficient and cause less environmental damage. *National Parks and Wildlife:* to ensure that the coverage of ecosystems is as complete as possible; and to investigate the ecology of some of the key plant and animal species. *Catchment studies:* to investigate ways to reduce erosion, and the adverse effects of logging operations on water quantity and quality, and on fisheries.

(d) *In education and training.* To expand education and training in forestry, timber harvesting, the forest industries and wildlife management. This should be carried out in collaboration with STIDC, the universities and the private sector. It is recommended that a model logging concession should be established as one of the training facilities. More and better training should be available and mandatory for people at all levels, including chainsaw operators, and skidder and tractor drivers.

(e) *In public relations.* To improve mutual communication between the Forest Department and the public. An assured future for sustainable forest management depends on more attention being paid to the views and participation of local communities, and on sensitive explanations to the public and to local communities about the local as well as the broader advantages to be derived from good forest management for timber production and the national and local benefits of National Parks and Wildlife Sanctuaries.

Recommendation A2 – Reducing the total area of the annual cut
This is the first of the two fields in which performance falls far short of the objectives set in the Forest Policy. The means by which the recommended reduction in harvesting is brought about will require careful consideration by the planning staff of the Department, so that any adverse effects on forest industries are minimized. Measures could include: using the harvesting of State land forests as a buffer; and trading off longer concessions for smaller annual *coupes*. These measures must be accompanied by intensified silvicultural management and by the rapid expansion of the PFE to its final size by the immediate gazetting of as much as possible of State land forest. A Commissioner should be appointed and a Unit established, specifically charged with accelerating and completing the process of gazettment, including that of the additional recommended TPAs.

Recommendation A3 – The improvement of the standards of catchment protection in the hill dipterocarp timber production forests
This is the second field in which implementation falls furthest behind the objectives set in the Forest Policy. Management must be improved by more precise prescriptions based on research, by withdrawal of logging from critical areas and by strict control of the conditions of licences.

B *For international cooperation and assistance*

The Mission found a strongly demonstrated commitment by the Governments of both Malaysia and Sarawak to sustainable management and conservation for Sarawak. In terms of assistance and cooperation, the Mission recommends the following to the ITTO and the international donor community:

Recommendation B1 – Manpower development
To assist the Government of Sarawak in establishing and implementing a continuing programme of manpower development so as to:

(a) deploy and apply the strengthened staff fully and effectively with the minimum delay;

(b) accelerate the rate at which staff acquire experience in both general management and specialized fields;

(c) improve the planning capacity of the Government for the development and review of the basic plan for the sector, and for providing the necessary supporting information; and

(d) raise the standards of technical and management skills at all levels in the logging and timber industries.

Recommendation B2 – Long-term outlook
To assist Sarawak in the preparation of the outlooks for long-term demand and supply which must underlie the initial basic plan for the sector under Recommendation A1 (b) above.

Recommendation B3 – International assistance
To raise, if the Government feels it necessary, additional manpower from international sources to assist the Sarawak Forest Department during the transitional phase and in the implementation of Recommendations A1–3.

Source: ITTO, 1990b.

in their eyes, would have provided a powerful springboard for further action. Some viewed the position of the NGOs as a brave stand on principle; others as an act of political naivety.[13]

The case of Malaysia and particularly Sarawak had wider implications. International criticism was building up in the late 1980s with the threats of logging bans and the public interest aroused by blockades of logging roads by the Penan. Malaysia was incensed because it rightly believed that, on the whole, its forest management was among the best practised by any tropical country. This reaction to criticism was almost certainly a cause of the extremely tough stance taken by Malaysia at UNCED, a stance that led to forests becoming such a contentious and adversarial issue. It is arguable that a different and more conciliatory approach might have brought better results in reforming abuses in Malaysian forestry and in furthering the international agenda on tropical forests. It took several years and myriad international meetings to rebuild the mutual trust and confidence that was squandered at Rio.

Part of this drama was played out in ITTO. WWF Malaysia found itself in a very difficult and embarrassing position in relation to its big brother WWF International. It had, for a number of years, worked closely on conservation projects with the Government of Sarawak, whose confidence it had gained by constructive criticism. The following is a quotation from a policy document prepared by WWF Malaysia in 1989:[14]

> If WWF wishes to remain in a position to support or influence policies and action in Sarawak in the interests of conservation, it must retain good links with the State and Federal Government. It should do so by continuing to give technical cooperation to the authorities. Such cooperation should be given without compromising the Fund's conservation goals.
>
> The acceptability of WWF's views to the Government is bound to be affected by the way in which those views are put forward. Arguments that are opposed to the Government's way of thinking are likely to get a hearing so long as they are courteously presented and well supported technically. It is, however, a fact of life in Malaysia that arguments that are presented through public channels (ie the media) are far less likely to get a hearing – and more likely to produce a counter-reaction – than the same arguments presented privately. The State and Federal Government would be very unlikely to accept WWF as a partner in conservation if it also played a role as a vocal public critic... The Fund should continue to emphasize its role as an independent, scientific and technical organization that cooperates with Government Agencies in the field of conservation; and all branches of WWF should avoid the role of high profile pressure group in relation to specifically Malaysian issues.
>
> If WWF is to continue to be active in Sarawak, it must work on specific projects as well as keeping up its general relationship with the Government. WWF should therefore keep up the present exercise of identifying specific, realistic goals for conservation in Sarawak and ... identify priorities for WWF funding and action.

WWF International did not take this advice.

A full presentation and discussion of the Sarawak report took place at the 9th Session of the Council in Yokohama in the autumn of 1990. In his introduction the leader of the Mission, the Earl of Cranbrook, commented that general public reactions to the report had been influenced to some extent by the fact that most opinions had been formed before the report had been read. It should not be judged on its recommendations alone; these were necessarily brief and selective. It should be discussed *as a whole*. The overall assessment of the Mission was: *'that sustainable forestry could be achieved, was being achieved in some respects but failed in others. The failures were remediable and recommendations were made to remedy them'*. And he emphasized that 'the dispatch of the Mission constituted a unique event and initiative, given the risks of the political sensitivities of governments. It was imperative to capitalize on this experience and build on this track record of performance'.[15]

The spokesman of the State Government of Sarawak spoke next, setting out the position of the authorities of Sarawak on the Mission's report. The Government 'accepted the report as fair and very constructive. They accepted the recommendations in principle and were taking measures to overcome the inadequacies and weaknesses in forest management as laid bare in the report.'

He gave a background description of the development strategy of the State. The timber industry and the forest sector provided not only substantial sources of employment to the people of Sarawak but also a major portion of the Government's revenues.[16] It was a matter of importance and priority to the Government that the rural populations benefited from these revenues.

He then raised some of the major issues discussed in the report. First, there was the State's position on the traditional rights and claims of the native communities in relation to forested lands. He stressed that these were 'matters for determination in the courts of Malaysia. Any aggrieved person could seek redress from the Courts of Law.' He quoted a recent instance where the High Court had dismissed, in June 1990, a case in which 'a group of natives from the Baram District, claiming native customary rights over forests, [had] filed a civil suit against the State and the timber companies involved'. 'The Sarawak Land Code did not recognize claims of native customary rights on land under forests. This fact and, generally, the land laws of the State were often misunderstood, hence the baseless and false claims which would lead to tribal conflicts if they were upheld.'

He went on to state that the Government of Sarawak accepted the goal of sustainable harvest as a cornerstone of its forestry policy and the recommendation to phase down the rate of harvest in the primary hill forest to a sustainable level. 'A drastic reduction in logging rate would, however, result in job losses, with consequences for the increasing incidence of shifting cultivation.' But he took issue with the conclusions in the Report about the current rate of harvesting and the future of primary forests. 'It was the Government's view that the forecasts were based on excessive estimates of the consequences of harvesting.'

In addition, the Government was taking steps to: (a) raise the area of PFE from 4.5 million to 6 million hectares; (b) raise the area of TPAs from 0.25 million hectares to 1 million; and (c) use industrial and other programmes to minimize the adverse effect on employment and loss of revenue when phasing down logging.

The environmental NGOs continued to use the report of the Sarawak Mission as a rod to beat the back of the Council.[17] Their spokesman 'deplored the lack of determination on the part of the organization to tackle issues central to its mandate'. He cited, as a case in point, the Council's decision on the report of the Mission to Sarawak:

> The Council's decision had failed to address several key issues raised in the Mission's report... [This] demonstrated that it had set aside the principles and standards enunciated in the ITTO Guidelines. What the Council had done instead was to approve a disparate package of projects, thus showing up the inadequacies of the internal mechanisms for project appraisal and funding. The attitude shown by the Organization over the question of Sarawak had cast doubts on the commitment of the members to make an effective response to the problems caused by logging.

A further statement on progress was made one year later, at the 11th Session of Council (November, 1991). Mr Hamid Bugo[18] spoke with special authority, being not only the delegate of Malaysia but also the Deputy State Secretary and Permanent Secretary of the Ministry of Resource Planning of Sarawak. 'Sarawak wanted to determine the extent to which its forest management was sustainable. The impetus ... was the genuine desire for rational utilization, not a reaction to external criticism.' He reported that TPAs covered 290,000ha and steps had already been taken to increase this to 1,000,000ha; the PFE totalled 4.5 million hectares and measures to increase this to 6 million hectares were substantially complete. 1.7 million hectares were 'State lands' and were being converted to agriculture. The total timber harvest in 1990 was 18.8 million m³, of which 12.5 million m³ came from the PFE but 6.3 million m³ from the State lands. He then made a curious point, which indicated a serious misunderstanding about the nature of sustainability.

> The central focus of the ITTO mission was sustainable management of the PFE (4.5 million ha increasing to 6 million ha). The report estimated the annual expected sustainable yield of this area at 9.2 million m³. As a lay person in forestry technical matters, it was baffling to the speaker that a set figure for sustainable yield could have been determined by the mission when the whole definition of the meaning of sustainability was still being hotly debated, even now, by the ITTC.

He continued:

> Nevertheless, Sarawak accepted the Mission's estimates in good faith, pending new information from growth and yield studies. The State has promised to reduce timber harvests from the PFE by 3 million m³ over the next two years, 1.5 million m³ during each of 1992 and 1993.

He went on to point out that, since the forestry sector contributed very significantly to State income, the reduction would mean a loss of employment and income, besides diminishing State Government revenues. This would mean accelerating growth in other sectors which might be of limited or late effect. He appealed to other members to assist by increasing prices paid for Sarawak's timber exports, supporting domestic processing, human resource development and liberalizing market access.[19]

Many delegates expressed their appreciation of the progress that was being made and wished to be kept informed, but Brazil once again reiterated its objection to the Mission. Mr Thomas Jalong, the representative of the Friends of the Earth, Malaysia, was more critical, as a member of a community group in Sarawak affected by logging. He rightly noted that the items mentioned by Mr Hamid Bugo – the increases in TPAs and in the PFE, and the reduction of cut – only complied with one part of the recommendations of the Mission. There was no mention of the reserve for local communities, cessation of logging in areas of disputed land claims and the value of non-timber products. He then criticized the Mission report. It gave a very narrow view of sustainable management, and land and access disputes were excluded from its recommendations. He called upon ITTO to admit indigenous peoples to the making of decisions. Some countries had already taken progressive steps to demarcate customary land rights. Would ITTO 'shy away from its responsibility to insist that its Member Governments adopt proper measures to recognize the rights of indigenous peoples living in tropical forests'?

The next and last report was given to the 19th Session (November 1995) by the Malaysian Minister of Primary Industry, Dr Lim.[20] The State had taken steps to implement almost all of the ITTO recommendations: log production from the PFE was now down to 9.2 million m^3; protected areas had increased from 300,000ha to 1.03 million; and the PFE had been expanded from 4.5 million hectares to 6.1 million. Dr Lim once again emphasized the cost: the loss of 26,000 jobs; more than US$100 million of lost income to workers; and losses of US$50 million in revenue and US$500 million in foreign exchange. He considered it very important that such efforts be formally recognized by Council and called on Council 'to evaluate the actions of member countries to enhance the management of their forests in accordance with the criteria and indicators and issue a written statement in recognition and acknowledgement of such efforts'.

One of the strong points of ITTO as an organization is that it is able within its own resources to move rapidly to transform recommendations and decisions into practical support. In this, it has an advantage over other multilateral aid organizations. In the case of Sarawak, it was able to do so by providing grants for four projects deemed by the State Government important in order to implement some of the more significant recommendations of the Mission: US$151,184 for manpower development of the Sarawak Forest Sector; US$1,677,198 for studies of the management standards of hill dipterocarp forests in Sarawak from a watershed management point of view; US$3,758,392 for the development of the Lanjak-Entimau Wildlife Sanctuary as a TPA; and US$462,577 for strategies for sustainable wood industries in Sarawak. In addition, there were a number of projects for Malaysia as a whole which undoubtedly produced some incidental benefit for Sarawak.

However, there has been one serious deficiency – there has been no evaluation. The Mission tried to formulate its recommendations in the light of the social and economic changes that might reasonably be expected in Sarawak. Since then, social development has moved apace and the economic climate has fluctuated (for example the economic downturn in East Asia and fluctuations in the market for tropical timber). It would be valuable to conduct a new assessment to determine what action has been taken on all the Mission's recommendations, what effect these have had on the practice of forestry in the field, and what have been their wider influence on the condition of the people of Sarawak and of their forests.

8

Ferment 1990–1992

'ITTO had consistently both presented itself as a unique forum for producers and consumers, and claimed itself a political success, but in the NGOs' opinion there was a disparity between the objectives and priorities of the ITTC and the reality of its achievements. Too much time was devoted to projects and not enough to policy'

Theo Anderson, Friends of the Earth

This was the beginning of a period of very rapid change in the international forestry scene. ITTO both influenced developments and was influenced by them. The most notable event was the United Nations Conference on Environment and Development (UNCED) held in Rio de Janeiro in 1992, known variously as UNCED or Rio 1992. The main outputs from this conference were the two Conventions – the Framework Convention on Climate Change (UNFCCC) and the Convention on Biological Diversity (CBD), Agenda 21 (of which, chapter 11 was entitled 'Combating deforestation'), and the Forest Principles. The last, in order to satisfy all the niceties of international negotiation, emerged with the title of 'Non-Legally Binding Authoritative Statement of Principles For a Global Consensus on the Management, Conservation and Sustainable Development of all Types of Forests'. For some time afterwards, it was essential in international meetings to use the whole phrase 'management, conservation and sustainable development of all types of forests', even when the question being discussed was much more limited in scope! But the importance of these documents cannot be underestimated, however contorted their language; for, like the Decisions of ITTO, they represent internationally negotiated commitments.

The great importance of UNCED was that it brought together, for the first time, many Heads of State to discuss 'the environment' and the future of the world's forests. The nature of the debate had changed, however, in several significant respects. It was, of course, inevitable – and necessary – that forests should be considered in the context of 'sustainable development', which after Brundtland had become, and has ever since remained, the Holy Grail of politicians. In fact, Agenda 21 and the Forest Principles did much, indirectly, to define what was meant by the phrase – and to ensure that social elements were given due weight in determining sustainability.

Over the preceding few years, a number of other burning issues had surfaced, among them: an emphasis on national sovereignty over natural resources; concern about the condition of temperate and boreal forests; the question of global versus national responsibilities in matters of the environment; and intellectual property rights in relation to living things and indigenous knowledge concerning them.

The emerging emphasis on 'all types of forests' – meaning temperate, arid-zone and boreal as well as the more emotive tropical – was first and foremost stimulated by irritation, particularly on the part of Malaysia and Brazil, at international criticism of their forest policies and the way in which it was thought that their indigenous peoples were being marginalized.[1] In the case of Malaysia, criticism was concentrated on the intensity of logging in Sabah and Sarawak and the plight of the Penan in Sarawak; in Brazil on the opening up of the Amazon with new highways, the generation of carbon dioxide by the burning of its forests, and the fate of indigenous tribes. Indonesia, too, was implicated because of the destructive exploitation of its forests for timber and the clearing of forests in Kalimantan to provide for the 'transmigration' of people from over-populated Java, Bali and Madura. But the mantra 'all types of forests' was also fuelled by concern about the way in which some temperate and boreal forests were suffering from pollution (especially in Eastern Europe) and by mining for timber in the USSR and British Columbia. One publication which was influential in widening the forest debate was the WWF work *Forests in Trouble: A Review of the Status of Temperate Forests Worldwide.*[2]

Brazil was the leader in raising the 'global versus national' controversy. Why should tropical countries sacrifice opportunities for development in order to retain forest, which was of importance to the whole world? Should not the global community pay for benefits forgone in the causes of preserving biological diversity, storing carbon and maintaining atmospheric oxygen? All of these questions (and others) portrayed with varying degrees of accuracy, were points of contention in the Rio negotiations. The environmental NGOs played an important part in raising issues and forcing the political pace.

It was in this climate of opinion that the ITTA came up for renewal and it is hardly surprising that the process of renegotiation was strongly influenced by the events surrounding UNCED – in particular whether the scope of the Agreement should be broadened to include trade in timber from 'all types of forests' or even be subsumed in an all-embracing forest convention.

The other main developments were concerned with the trade. One involved its depressed prospects. The other revolved round the ultimate effect of international measures affecting trade in timber (bans, tariffs, certification) on the quality of management, first, of the forests from which that timber was harvested and, more widely, of all tropical forests.

FAO published a Global Forest Resource Assessment for the year 1990. The TFAP, inaugurated at the World Forestry Congress in 1985, was gathering momentum – and criticism. In its original form, it had proved to be too much of a gathering ground for proposals for traditional forestry projects and it was criticized for its lack of procedures for wide consultation and for its tendency to focus on the forestry sector rather than on the effect of wider policies on the condition of the forest. In 1990, the name was changed to the Tropical Forests Action Programme to reflect changes in its emphasis. During 1991, there was a 'High-level Independent Review' and, thereafter, a series of meetings to discuss various future options, until, at the 103rd Session of the FAO Council in 1993, it was finally decided to set up a Consultative Group on the Tropical Forests Action Programme (TFAP-CG) with a wide membership.[3]

ITTO took part in the TFAP Advisors Group and ITTO projects were planned to be complementary to those of other funding agencies, as integral parts of the national plans being developed under the TFAP umbrella.[4]

After UNCED, ITTO played a critical role in launching the Intergovernmental Panel on Forests (IPF) by acting as intermediary for the Government of Japan in providing US$1 million to establish the IPF Trust Fund.[5] This enabled IPF to recruit its Coordinator and Head of the Secretariat, Dr Jagmohan Maini, and a senior policy advisor. It worked closely with IPF and the subsequent Intergovernmental Forum on Forests (IFF), helped with a further secondment, became part of the Inter-Agency Task Force set up in association with IPF/IFF and is now (2002) part of the Task Force's successor, the Collaborative Partnership on Forests (CPF).

The year 1990 saw the beginning of a period of great activity for ITTO. This and succeeding chapters describe some of its accomplishments in the policy arena and the difficulties encountered. It deals with ITTO's achievements in establishing norms (through its guidelines, criteria and indicators, and its country missions); in implementing the Year 2000 Objective; and in the development of its action plans. It also covers the renegotiation of the Agreement, market issues, the interaction with the Convention on International Trade in Endangered Species of Wild Fauna and Flora (CITES), and the declining input of environmental NGOs.

International Tropical Timber Organization Action Plan 1990

In 1989, the Council had advocated that a single action plan should be prepared based upon the action plans of the three committees and this, sub-titled 'Criteria and priority areas for programme development and project work', was prepared and approved in 1990.[6] It identified three special features of ITTO in the international community: the combined objectives of utilization and conservation; the equal partnership between producing and consuming countries; and its unique promotion and facilitation of inputs from NGOs, industry and trade. After some general paragraphs identifying the special roles of ITTO in the international arena and the basic principles governing its operation, it set out strategies and action plans for the three segments of its work (reforestation and forest management, forest industry, and economic information and market intelligence). Finally, there were conclusions, and appendices covering 'problems and opportunities' and the elements of the work programme. This Action Plan at last gave some guidance on the fields that

Table 8.1 *Problems and opportunities*

Problems	Opportunities
The tropical forest base and its resources are in the process of accelerating depletion and degradation.	Integrated national and international actions aimed at arresting the decline and degradation of the tropical forests.
The prospects of sustainability differ.	Each country must determine its own strategy.
Long-term land use and forest policies are mostly non-existent or not adequately applied.	Coordinating ITTO activities in land use and forest planning activities.
Lack of knowledge hampers sustainable forest development.	Organization of existing knowledge and development of human resources in the field of sustainable forest management can be initiated by immediate actions. R&D programmes and projects can be initiated after thorough problem formulation and preparation.
Sustainable supply of wood raw material at competitive prices is the most critical challenge facing the tropical timber industry.	Scope for improved or sustained supply.
Export of logs is still dominating – the industry is weak.	Scope for improvement of the forest industry.
Lack of finance is also a problem.	Financing depends on the prospects.
The tropical timber market lacks transparency.	Scope for improvement of the tropical timber economy.
Market information – a problem of cooperation, coordination and documentation – a problem of willingness and capability to provide information that is reliable and timely on a regular basis.	Increased ITTO in-house capacity is a key to market information services to the ITTO members.

ITTO would, or would not, support. It tended, however, to include everything and gave little indication of priorities. Table 8.1 shows the problem analysis that accompanied the strategy mapped out by the three committees.

The Quito Round Table and the Year 2000 Target

The 10th Session of the Council, held in Quito, Ecuador, in May 1991, was very important for ITTO. At the Session, a two-day round table was held on the subject 'The Agenda for Trade in Tropical Timber from Sustainably Managed Forests by the Year 2000', based on a pre-project report commissioned from the Oxford Forestry Institute[7] and on the *ITTO Guidelines for the Sustainable Management of Natural Tropical Forests*. In the long term, this round table was viewed as a 'start in evolving practical policies to strike a rational balance between conservation and utilization for tropical forest resources; in the short term, the debate … would be ITTO's contribution to the UNCED process, in particular its Agenda 21'.[8] Presentations were made by the producer and consumer countries, by intergovernmental organizations and by NGOs representing conservation, the timber trade and industry.

The record of the discussions gives a useful insight into the preoccupations of the time and the significant emphasis on incentives.[9]

The objectives [of forest management] were defined as the stabilization of the forest estate, maintenance and enhancement of resource management, and optimization of forest resource utilization. To meet these objectives … incentives might be directed towards institutions, towards commercial organizations and towards the needs of local communities (social needs). Within this context, both financial and non-financial incentives could be considered.

The overall conclusions [of the round table] were clear: (a) A tailor-made package of incentives would be required for each recipient country that should be closely linked to the declared objectives of the national forest policy; (b) Progress of forestry development should be reported in forest management terms; and (c) Transfer of resources, although recognized as a universal parameter, would be conditional on the particular circumstances of recipient countries.

There was unanimous support for programmes to ensure the sustainable management of tropical forests, and trade in tropical timber from sustainably managed resources by the year 2000. The ITTO Guidelines for Sustainable Management of Natural Tropical Forests were universally accepted as a suitable framework for all countries to develop their own national guidelines to meet the requirements of Sustainability 2000.[10]

There was agreement that the causes of forest destruction were multi-faceted and a unique set of imperatives would stimulate each country towards sustainable forest management programmes. Reality [required] that each sovereign nation should prepare a national action plan…

The pragmatic concept of Sustainability 2000 [was] based on the production of defined goods and services with an ecosystem where all functions were maintained. It was recognized that not all forests needed to be sustained for all goods and services as long as a defined national balance was maintained.

In order to substantiate progress towards sustainability, a regular reporting procedure was considered essential. Among those discussed were: (a) continuous assessment of progress within the national action plan; (b) regular assessment of international trade in tropical timber products; (c) certification systems which confirm progress in sustainable forest management practices.

The position of ITTO as a pivotal organization in the establishment and implementation of national plans was emphasized…

It was generally agreed that incentives were required to achieve the objectives of Sustainability 2000. Care was needed in the application of incentives to ensure their focus on the declared objectives of sovereign nations' needs...

Forest management practices within concessions could be made accountable by extending the concession period over a time scale which guaranteed the sustainability of the forest...

Enhancing the value of standing timber might be accomplished (a) by means of a concession auction system (including a reserve price); (b) through linkage with FOB prices for the wood product; and (c) by offering incentives to buffer stumpage prices in times of market downturn...

Consumer countries, in expressing their commitment to Sustainability 2000, stressed their determination to support the liberalized trade in tropical timber within the framework of the multilateral trading system. Disincentives to free trade were identified as the imposition of import tariffs and export taxes, unilateral actions which restricted sales in consumer countries and unsubstantiated and inappropriate certification schemes.

There was nothing particularly new in these sentiments but they did crystallize a developing consensus within ITTO members; there followed a number of conclusions which transformed these points into recommendations – that ITTO should encourage, invite, initiate, devise, support, introduce, facilitate and explore various actions. These were then translated into the terms of the highly important Decision 3(X) (see Box 8.1).[11] This Decision set the course for much of the action of ITTO in succeeding years.

Other Issues at Quito

Various other issues were raised in Quito – Sarawak, trade restrictions and, most significant, the attitude of the NGOs to developments within ITTO. Progress was continuing on guidelines for plantations and for biodiversity conservation. Some valuable ideas raised in this session were also to contribute to the development of future policies.

Trade

Mr Alle Stoit spoke on behalf of UCBT of Europe, the International Wood Products Association (IHPA) of the US and the Japan Lumber Importers' Association (JLIA), together serving a market of 700–800 million consumers. The importing trade was concerned about supplies of tropical timber in the long term, while there was resistance among consumers who were anxious about the future of the forests. The trade provided a large income for many people in developing countries. The trade was attempting to inform customers of the complex social and economic problems facing these countries and the steps they were taking towards sustainability. Refusal to buy tropical timber was only

Box 8.1 DECISION 3(X): SUSTAINABLE TROPICAL FOREST MANAGEMENT AND TRADE IN TROPICAL TIMBER PRODUCTS

DECISION 3(X)

Decides to adopt and implement the following Strategy by which, through international collaboration and national policies and programmes, ITTO members will progress towards achieving sustainable management of tropical forests and trade in tropical timber from sustainably managed forests by the year 2000.

1 In developing and implementing this Strategy, Members are invited to:
   (a) Continue to develop forest practices and regulations for sustainable management of the tropical forests taking into account the ITTO Guidelines, national and local conditions;
   (b) Provide to the XI Council Session a paper on their proposed progress towards the Year 2000 target;
   (c) Enhance their ability to attain the Year 2000 Target by investigating liberalised trade in tropical timber within the framework of the multilateral trading system;
   (d) Confer annually on the progress towards the Year 2000 Target;
   (e) Support projects leading to the achievement of sustainability and the Year 2000 Target through the Special Account;
   (f) Inform the ITTO by June 1995 through reviews of progress made towards achieving the Year 2000 Target;
   (g) Facilitate the flow of technology that will improve the management of forests, the utilisation of tropical timber and value of timber products;
   (h) Consider available studies on incentives.
2 Pursuant to this Strategy, the ITTC will:
   (a) Encourage national strategies which include, among others:
      (i) Forest conservation and management;
      (ii) Appropriate economic policies for forest and timber, for example full cost forest accounting and resource pricing regimes;
      (iii) Identifying incentive schemes for attaining sustainable management of tropical forests and regarding operational steps to this effect as ITTO activities;
      (iv) The investment of revenues from forests in sustainable forest management; regeneration; and expansion of the forest estate through plantation development;
      (v) The enhancement of the ability of local communities, particularly those within or near the forest, to obtain appropriate returns and other benefits from sustainably managed forests;
   (b) Undertake in 1995 a major review of progress towards the achievement of the Year 2000 Target, based on national submissions based on 1(f);
   (c) Suggest, in the light of the 1995 review, any further measures for attaining the Year 2000 Target;
   (d) Consider, at its XI Session, the conclusions and recommendations of an expert panel convened by the Executive Director to develop methods of defining and measuring sustainable tropical forest management and decide on any further initiatives to ensure the attainment of the Year 2000 Target;
   (e) Convene a workshop on incentives to promote sustainable development of tropical forest as provided for in PD 82/90 (M) which will identify options for

members in formulating trade policies and will be based on the guiding principles listed below:

(i)   Monitoring by members of the international timber trade through timely and accurate information on the market, including prices;

(ii) Improved utilization and increased value of forest products through the introduction of improved technology and management practices, making more efficient and better use of tropical forests and forest products;

(iii) Market mechanisms to cover the costs of sustainable management of tropical forests;

(f) Undertake assessments of the resources needed by producer countries to attain sustainable management of tropical forests by the Year 2000 Target. This will be achieved through a systematic approach which will identify the obstacles to and the solutions for attaining sustainable forest management and the time-stream of costs and benefits in implementation of the Year 2000 Target. This will be co-ordinated by a panel of experts selected by the Council. The terms of reference and composition of this panel will be agreed at the XI Council Session.

3 In support of the implementation of the Council Strategy, the Executive Director is requested:

(a) To identify, with assistance from an expert panel, possible methods of defining general criteria for a measurement of sustainable tropical forest management and present them to the XI Council Session;

(b) To undertake, with the assistance of an expert panel, studies to estimate the resources needed by producer countries to attain sustainability by the Year 2000 and to report on progress at each Council Session;

(c) To assist member nations, that so request, in undertaking studies and projects which define the most appropriate incentives for sustainable forest management, taking account of the ITTO Guidelines, and trade in timber from sustainably managed resources and to report their findings and progress to Council;

(d) To work with independent organizations, including industry associations and other non-governmental organizations, in developing means by which they can help Members to achieve the Year 2000 Target; and

(e) To explore with other relevant international agencies the possibility of a joint study of the sale and pricing systems for standing timber and their contribution towards sustainable management of tropical forests.

a gesture, as only a small proportion entered international trade; most was used domestically and domestic use was on the increase. Priority should be given to ITTO activities which 'directly aided producer countries in their supervision and management of forest';[12] progress should not be hampered by arguments about the definition of sustainability. Mr Stoit said that he believed that legitimate national land use plans would lead to the loss of 0.5–1 billion m^3 of tropical timber through land clearance. He questioned whether this timber would qualify as sustainably produced.

Mr Arthur Morrell spoke as representative of the trade and industry organizations. Traders from around the world had agreed to work together in contrast to the fragmentary approach of the past. The task was to ensure continuity of progress towards the Year 2000 Objective. He referred to the contribution of the producers and their concern about interference by external agencies in domestic policy matters. Few developed countries would tolerate such interference; partnership was preferable to policing.

The trade, he said, was completely opposed to any 'labelling' that would exclude timber from legitimate conversion forests; such timber had a proper place in satisfying demand. The trade would support a sound system of certification introduced at the correct time. A good start would be for producers to adopt the ITTO Guidelines for the Sustainable Management of Natural Tropical Forests and to commence working on their own national versions. The importing trade, and not just environmental NGOs, had a right to be considered lovers of forests. The two held common objectives and could cooperate; the trade would play its part.

NGO attitudes

There were three strong statements from environmental NGOs, each giving a different insight. Mr Evaristo Nukuag, speaking for COICA (Coordinadora de las Organizaciónes Indigenas de la Cuenca Amazónica) on behalf of five indigenous peoples' organizations from Brazil, Bolivia, Ecuador, Colombia and Peru, argued that the people most concerned about the future of the tropical forests were those indigenous to the area; he questioned whether any of his audience would permit strangers to make decisions about the future of those living in their homes. These people, he said, were aware of what was happening and had made common cause with environmentalists in developed countries. They had their own techniques of 'sustainable management'; for them, human and land rights were one and the same thing. Human beings were part of the Amazon biosphere and account must be taken of their welfare. Mr Nukuag questioned whether the money spent under the TFAP had been well spent. Indigenous people were not opposed to development but they wished to propose alternatives for the development of the Amazon which took into account their traditional rights. He asked ITTO to initiate a study into the effects of logging on the lives of indigenous peoples.

The increasing impatience of the environmental and conservation organizations was articulated by Mr Chris Elliott of WWF, speaking on their behalf. The WWF had been faithful in its attendance at Council meetings since the beginning, fielding as many as 13 representatives at the Bali meeting. He emphasized that the intention of the organizations he represented was to be constructive but they wished 'nevertheless to express their concern that ITTO's activities were not on a scale commensurate with the problem facing tropical forests'. He criticized 'the tendency for fruitless discussions, bargaining on projects concluded without transparency and the perceived lack of commitment of many delegations to the ITTA, 1983, the Action Plan and Target 2000 [Year 2000 Objective]. The producing countries' statement in Bali had been forgotten.' He urged members to 'really develop a strategy … for Members' progress to Target 2000, *quantified throughout with dates for implementation of guidelines, areas of forests being brought under sustainable management and volumes of sustainably produced timber*.'[13]

Policy development

Other significant developments were brewing in the Committee for Economic Information and Market Intelligence; three project ideas were submitted by the UK, the first of which, in particular, was destined to have an important influence on future policies and priorities. All three were subsequently implemented.[14] They were:

- The economic linkages between the international trade in tropical timber and the sustainable management of tropical forests.
- Study of the feasibility of strengthening incentives for the management of tropical moist forest for sustained production in a producer country.
- Forest accounting: monitoring forest condition and management.

The objectives of all three were closely linked. The first was fundamental to the whole rationale of the Organization; the second was a study of the means by which sustainable management might be brought about; and the third was a mechanism for monitoring progress. All three were designed to provide vital underpinning to the Year 2000 Objective.

It was agreed to move ahead on all of these.

The 11th Session of Council – Yokohama, November 1991

In his introductory statement, the Executive Director summarized many of the issues facing tropical forests in the build-up towards UNCED: developing countries feared 'that the industrial nations of the North, having depleted their own environmental capital for their own development, were now attempting to rebuild it by placing environmental curbs on the South's development.'[15] They were of the view that 'the North should assume certain obligations to make up for its past misuse of the planet's ecological capital' and that it was the 'North's duty to aid with resources to protect the environment and to generate economic growth. Socially necessary growth, plus environmental protection, was essential to ensure reasonable employment and basic goals in health, education and poverty alleviation.' He questioned whether the so-called world economic order was creating the conditions in which it was possible to protect the forest. He listed the fundamental causes of tropical deforestation – poverty, population growth, subsistence agriculture, land tenure and credit systems, lack of financial resources and investment, the debt burden, imbalances in community trade and threatening trade barriers: all of these called for action.

> The debate would rage with greater passion as long as the developed countries [appeared] to the developing countries as shirking the issue of new and additional financial resources needed to address some of the crucial underlying factors of the environmental dilemma.

The donors' response to the plea for additional resources was an insistence that the projects should be of higher quality, that there should be a better balance between project and non-project work, and that resources should be expended effectively and in the field, not on bureaucracy.

Progress towards the Year 2000 Objective

At this session, several countries presented progress reports towards the Year 2000 Objective: among the producers, Ghana, Indonesia and Malaysia; among the consumers, the European Community, Australia, Austria, Denmark, Finland, France, Germany, Japan, the Netherlands, Norway, Sweden, Switzerland and the UK. A number of members called for more to be presented before the next session.

The report from the delegate of the Netherlands, introducing his 'country document', covered ground that was later, in the 15th Session, to prove explosive.

> Chapters 5 and 6 summarized the objectives, encouraging tropical forest preservation through balanced and sustainable land and forest use with a view to halting the present trends in deforestation and environmental degradation. His Government would also encourage measures to bring harvesting into line with sustainable management and measures to completely prohibit harvesting in virgin forest, and from endangered tree species. The Government of the Netherlands believed that, from 1995, the tropical timber trade should be restricted to countries with policies geared to sustainable timber production and protection.

The 12th and 13th Sessions of Council – Yaounde and Yokohama, May and November 1992

Impatience

By the time of the 12th Session in Yaounde in May 1992, the impatience of the NGOs was manifest. In what was becoming a standard refrain, Mr Theo Anderson (Friends of the Earth, Ghana), speaking on behalf of the environmental NGOs, expressed disappointment about the lack of progress towards the Year 2000 Objective.[16] He criticized the lack of a strategy, the absence of agreement on reporting progress and the ambiguity of the definition of the Objective itself, the failure of ITTO to allow an independent review of its activities, its lack of commitment to the Agreement and its reluctance to listen to the NGOs' point of view.

> NGOs had supported the Organization's ideals because they too shared the common objective … ITTO had consistently both presented itself as a unique forum for producers and consumers, and claimed itself a political success, but in the NGOs' opinion there was a disparity between the objectives and priorities of the ITTC and the reality of its achievements. Too much time was devoted to projects and not enough to policy.

His last remarks took the form of a veiled threat – the first hint that the NGOs might withdraw.

> NGOs were closely watching the renegotiation exercise and considered that it was not the only option to advance their ideals, or to reflect the concern inherent in their several criticisms of the Organization.

Mr Arthur Morrell, the representative of the trade organizations, also expressed impatience at the lack of progress. He reaffirmed:

> the timber trade's commitment to ITTO's sustainability objectives. Each Council session was an opportunity to advance this ideal and, although some progress had been made this time, in the trade's view it was insufficient. National delegations should include more trade representatives; these … could see at first hand the work of ITTO … the trade believed strongly in cooperation, but that partnership required a similar willingness on the part of other groups.

Trade

Timber restrictions were at the forefront of attention and there were increasingly acrimonious exchanges. The Executive Director was very outspoken. He drew attention to the serious concern expressed by many members over the campaign advocating restrictions or total bans on the import of tropical timber. The provisions of Article 30 of the ITTA, 1983 required members to avoid actions which would have the effect of running counter to the Agreement. Actions ranged from bans on tropical timber by city authorities to national legislation imposing unilateral discriminatory regulations 'euphemistically termed ecolabelling'. He commended the work of trade associations on their efforts to educate the public about the harmful effects of such legislation, mentioning the CURE programme of the IHPA, the work of the Tropical Forest Foundation (TFF), the 'Forests for Ever: Think Wood' campaign of the UK Timber Trade Federation and many others in Africa, Europe and Japan. Some producer nations had despatched missions to consumer countries to explain what they were doing and their trade missions were active in informing their customers. In spite of the energetic cooperation between consumers and producers, 'how was it possible that some countries had permitted their national legislatures to accept negative viewpoints as a basis for legislation'? For the governments of producer countries, it was the ultimate irony that, after incurring the displeasure of their own electorates for the imposition of restrictions on timber harvesting and uncontrolled conversion of forest to agricultural land, they were now threatened with trade sanctions imposed by the importers of their products. Dr Freezailah called on ITTO to 'act against these negative trends'.

The spokesman for the producers, in his speech, singled out Austria for particular criticism. Austria had recently passed a Federal Act on the Marking of Tropical Timber and Products of Tropical Timber, and on a Quality Mark for Tropical Timber and Timber Products from Sustainable Forest Management which came into effect on 1 September 1992. This required the mandatory marking of any product containing timber from one of at least 50 tropical species and a quality mark to prove that it originated from sustainably managed forest.

He reported that the producers' group was united in its opposition to this legislation.

> The grounds of their objection were, firstly, that a country endowed with temperate forests producing 14.6 million m^3 harvest in 1990 and earning 8% of its export earnings from forest products should not seek unfairly to enhance its competitiveness in third country import markets vis à vis tropical timber by setting an example of discriminatory legislation which ran counter to efforts to expand free trade. Secondly, the mere act of marking constituted a stigma which subtly persuaded consumers to switch to temperate timber, not to mention the additional costs of compliance with the quality mark legislation. The legislation … was a clear example of a non-tariff barrier … and totally inconsistent with the Statement on Forests to which Austria was a party at UNCED.

He carried the war into the opposite camp:

> Why had Austria not extended the legislation to cover temperate timber? Was Austria free of problems in sustainably managing her own forests? Did she not have acid rain, encroachment by grazing and excessive extension of the road network? Few tropical countries had so little residual natural forests as had European countries. Selective felling in tropical forests was criticized but clear-felling in temperate forests was not… Although there were earlier examples of local authorities taking such unilateral actions, Austria was the first case of a sovereign nation taking this approach. By impeding trade unfairly, Austria would contribute to a fall in demand and prices for tropical timber, a derived fall in value of forests, and a consequent increase in forest clearance for agriculture.

He appealed to Austria to repeal the legislation and was backed in this by Brazil, Indonesia and Cameroon. In the further debate, the representative of the European Union (EU) noted that the issue of labelling had been raised in discussions at the General Agreement on Tariffs and Trade (GATT), the European Community (EC) and the Association of Southeast Asian Nations (ASEAN), and contended that the subject needed a comprehensive technical review. Mr Simon Counsell, for FoE, made the point that debate on the issue was long overdue and that there were already several labelling and certification systems, all developed outside ITTO and none referring to the Organization's guidelines. (It is not clear whether this last point is praise or criticism of the labelling systems or of the guidelines, or a dig at the Organization's relevance!) Mr Counsell commented, with some justification, that the Council had been given the chance to debate the issue in 1989, when a labelling study had been brought before it by the FoE representative as part of the UK delegation.

The Delegate for Austria, Mrs Elizabeth Weghofer, explained her country's position. Public opinion had been calling for visible action on the conservation of tropical rain forests and the Austrian Parliament felt bound to acknowledge these concerns. The legislation was not motivated by protectionism; no quota restrictions were being placed on imports from any country. The measure affected equally the Austrian processors of imported timber. It was particularly designed to inform a consumer market increasingly demanding better information. Since Austria recognized that timber was an important export for some countries, she was willing to intensify dialogue; a high-ranking delegation had recently met members of the Malaysian Government. Austria supported ITTO's Year 2000 Objective and its work on guidelines and criteria and had recently allocated 200 million schillings for assistance in sustainable forest management. Finally, she 'regretted the defensive reflexes apparent in the Executive Director's address but welcomed multilateral discussions on the labelling issue within the framework of ITTO'.

Projects and their funding

Meanwhile the processes of filtering, revising, approving and funding projects continued. The 13th Session may be taken as an example. It will be remembered that projects are funded from the Special Account; pledges, especially unearmarked pledges, to the Special Account are therefore of particular value.

In this session, Japanese industrial members pledged 10 million yen for conservation and sustainable management and another 10 million yen for sustainable development of tropical forests; the US pledged US$1 million 'in recognition of continued improvements in the project cycle and in the quality of project proposals'; Switzerland announced US$1 million 'on projects that would help the attainment of the Year 2000 Objective'; the Netherlands pledged DFl 120,000 for the same purpose; and cooperative remarks without any actual commitment were made by the UK, Denmark, Norway, Australia and Sweden.[17]

At the same session, 19 projects were approved to the value of US$14,612,579 (11 of these had been revised once, two twice and two three times). Of the 19, five were funded, totalling US$2,388,879. Funding was also authorized for one previously approved project to the extent of US$556,239. Funding was authorized for the remaining 14 projects approved in this session (US$12,223,700) and US$5,899,759 for projects approved at previous meetings 'as soon as earmarked funds are available'. It is evident from this that substantial numbers of projects were funded, but that pledges were totally inadequate to fund all approved projects even though these had been through an improved and now rigorous selection process.

Attitudes

The NGOs and the delegate of the USA, Mr Robert Johnston, had the last word at this session.

Mr Simon Counsell, thus:

> The Session was overwhelmingly dominated by debate on projects to the exclusion of policy reforms. Terms of trade, market access, and other central issues were neglected while consumer delegations diverted attention to projects. Consumer delegations had failed to state how they would halt the import of products from illegally felled trees and help producers to prevent such criminal activities. Whenever ITTC had discussed policy issues, as with labelling, their approach had been reactive and acrimonious rather than proactive and cooperative. If the meetings had shown anything, it had shown how great was the need for a better successor agreement based on trust instead of double standards... The NGOs commended the producers in looking for more radical alternatives for the new agreement.

Mr Robert Johnston of the US, speaking on behalf of the timber trade, reported that all involved in the trade, both producer and consumer, agreed that unilateral government-mandated labelling schemes harmed the trade. They lessened the likelihood that resources for sustainable management would be available to those producers who most needed them.

Yet it was not until a year later that the Council decided to set up a consultancy and working party on certification of all timber and timber products.[18]

Tropical forests, or all forests?
Renegotiating the ITTA

'We are sick and tired of the application of double standards; one for tropical timber and none for temperate and other timbers, and we want this eliminated under the Successor Agreement'
Producer Spokesperson

The years 1992 to 1994 were dominated by two events: in ITTO by the renegotiation of the ITTA; and in the wider international arena by UNCED (also known as the Earth Summit) and its aftermath – and many of the forest issues intermingled. UNCED took place in June 1992. The process of renegotiation followed almost immediately afterwards and lasted from early autumn 1992 until January 1994. During the whole of this period, the normal activities of ITTO continued, as we shall see in later chapters.

Forest issues were among the most contentious covered at the Earth Summit.[1] Arguments were polarized, principally along a North–South divide. During the Preparatory Committees, much of the North had argued that forests, although situated in national territories, were of global importance, principally for their biodiversity and their functions of climate regulation. They argued that a degree of supranational control of forests was desirable, and many proposed a convention, which would be legally binding.

The South, however, had stressed the sovereign right of countries to use forests for their development. They argued that global notions of sustainability could not encompass these varied and legitimate needs, but rather that more effective national and local control of forests was needed. The South recognized that forests did provide global benefits, but emphasized that, if there were to be a convention (and many in the South would not entertain this idea), this would have implications for sovereignty; thus a compensation mechanism would be essential to cover the loss of opportunities for development in return for the 'global forest services' rendered. When it became clear that a compensation mechanism would not be forthcoming, nearly all vestiges of support for a convention disappeared in the countries of the South. Rather, their argument became that the North was seeking a forest convention as the cheapest way of obviating the need to cut its own carbon emissions.

By the time of the Earth Summit, most countries had realized that there was little scope, and no time, to negotiate any form of legally binding forest instrument, although they were still divided over the need for one. Rather than risking a serious confrontation over a possible convention, most sought the positive approach of striving for a degree of consensus over forests.

The result was the Forest Principles – the *Non-Legally Binding Authoritative Statement of Principles For a Global Consensus on the Management, Conservation and Development of All Types of Forests*. The Principles did indeed, as stated in their preamble, 'reflect a first global consensus on forests'. One of their most significant aspects was in the process of their negotiation – no country walked out, in spite of the fact that not all agreed with every paragraph. This suggested a recognition of the global importance of forests, and a willingness to recognize demands that did not necessarily meet with national desires.

Very briefly, the Principles cover all forest types and many associated issues of environment and development. As stated in the preamble, the 'guiding principle is to contribute to the management, conservation and sustainable development of forests, and to provide for their multiple and complementary function and uses'. The importance of local, forest peoples and of women is recognized, as is the need to support their economic stake in forest use. They note the need for valuing forests, setting associated standards, monitoring them, using environmental impact assessment (EIA) for forest developments, setting aside protected forests, developing plantations, developing national and international institutions, and encouraging public participation. The document is also full of references to the need for additional financing and statements emphasizing the importance of sovereign use of forests.

The Principles have all the characteristics of an internationally negotiated text. They are involved, opaque and full of ambiguous and evasive phrases. Nevertheless, they have laid down the principles underlying an agenda for the future of the world's forests which has gained international currency and authority. They constitute, of course, a political document and should in no sense be read as operational; but they contain, nevertheless, many handles for positive action which were further developed in Chapter 11 of Agenda 21 and are the basis of international action to this day.

Chapter 11 of Agenda 21,[2] on the other hand, was exactly what its title implied – a detailed agenda for action, divided into four programme areas

- Sustaining the multiple roles and functions of all types of forests, forest lands and woodlands;
- Enhancing the protection, sustainable management and conservation of all forests, and the greening of degraded areas, through forest rehabilitation, afforestation, reforestation and other rehabilitative means;
- Promoting efficient utilization and assessment to recover the full valuation of the goods and services provided by forests, forest lands and woodlands; and
- Establishing and/or strengthening capacities for the planning, assessment and systematic observations of forest and related programmes, projects and activities, including commercial trade and processes.

This encompassed a huge complex of desirable actions and was accompanied by an approximate estimate of total costs. The average annual amounts for these four items over the period 1993–2000 were estimated to be as follows (amounts from the international community on grant or concessional terms in brackets): US$2.5 billion (US$860 million); US$10 billion (US$3.7 billion); US$18 billion (US$880 million); and US$750 million (US$530 million), respectively. It was suggested that funds should be directed towards strengthening the capacity of ITTO (among others) in the last two programme areas. The main implications for ITTO of these two documents were in the setting of standards, harmonization of trade, monitoring of forest condition and trade flows, national capacity building, provision of information and providing a forum for international discussion on reconciling the interests of North and South, and of claims of environment versus development (conservation versus use). In fact, business as usual but preferably more of it!

International Tropical Timber Agreement, 1994

The renegotiation of the ITTA took well over a year; a timetable is given in Table 9.1. An informal working group was held in Washington in September 1992; two meetings of Preparatory Committees (PrepComs) in November 1992 and January 1993. Finally the United Nations Conference for the Negotiation of a Successor Agreement to the International Tropical Timber Agreement, 1983 met in April, June and October 1993 and finally emerged with an agreement in January 1994. Meanwhile, the normal business of the Council continued – the 14th Session in May and the 15th in November 1993 – but there were also two special sessions, the first after the meeting of the Second PrepCom (January 1993) and the second immediately after the last meeting of the Negotiating Conference (January 1994).

Discussion started in earnest in the Informal Working Group on ITTA Renegotiation held in September 1992.[3] The meeting was very useful in laying the groundwork for the more formal meetings that followed. It was attended by people from consumer and producer countries, the Council and the ITTO Secretariat. It was made clear that each took part in his or her personal capacity; no opinions were attributable and no public statement was issued; its report presented a set of 'issues' and 'options' and did not necessarily represent a consensus.

The international context within which ITTO was operating was changing. It was now set by UNCTAD VIII and its 'Cartagena Commitment', by the forthcoming Uruguay Round and by the recent deliberations at UNCED, a very different set of circumstances from those surrounding the negotiation of the ITTA, 1983.

To start with, there were four different views about the possible form and scope of the future ITTA.

- First, there were those who considered that the Agreement *should not be changed*. It had the flexibility and internal consistency to enable it to cope

Table 9.1 *Timetable of the renegotiation of the International Tropical Timber Agreement*

UNCED	June 1992
Informal Working Group, Washington	24–25 September 1992
PrepCom 1	11–13 & 23–24 November 1992
ITTC 13th Session	16–21 & 24 November 1992
PrepCom 2	22–30 January 1993
ITTC Special Session 1	30 January 1993
Negotiating Conference Part 1	13–16 April 1993
ITTC 14th Session	11–14 May 1993
Negotiating Conference Part 2	21–25 June 1993
Negotiating Conference Part 3	4–15 October 1993
ITTC 15th Session	10–17 November 1993
Negotiating Conference Part 4	10–26 January 1994
Formal Statement by Consumer Members	21 January 1994
ITTC Special Session 2	22 January 1994
Statement by European Union	24 January 1994

with evolving circumstances. Future emphasis should be on improving the operational capacity, performance and impact. It should remain a commodity agreement with a focus on timber as a commodity while retaining its present environmental provisions for conserving the resource base through sustainable management.

• A second group thought that the Agreement *should be extended to cover all tropical wood and tropical forest products*, including tropical conifers and non-timber forest products, in order to include all those forest products whose management and extraction would have a direct effect on the management and sustainable development of the tropical forest estate.

• Others considered that the Agreement should cover *trade in all timber*. This would be in accordance with the holistic approach of UNCED – to take a global view of all forests and consequently of all timber. This proceeded from the assumption that it was only by encompassing all trade in timber and timber products that it would be possible to create the right environment to promote sustainable forest management and development.

• The fourth view was that the Agreement should foreshadow movement towards the conclusion of a *global forest convention*.

There appeared to be a consensus on a mix of options: (a) that the focus of ITTO should be on all tropical timber including conifers but excluding pulpwood; (b) that its mandate should be strengthened to promote sustainable forest development by encompassing the Guidelines, Target 2000,[4] the Forest Principles and Agenda 21; (c) that its normative role should be emphasized in promoting appropriate national policies and guidelines; and (d) that it should be strengthened so that it could provide a better range and balance of services, eg better trade and economic information.

The discussion ranged over all the clauses in the Agreement but was most active in considering the future goals and objectives of ITTO. While some considered that the existing wording was sufficiently general to be left unchanged, others were of the view that the changing circumstances required significant revisions and explicit statements on certain themes. Most thought that sub-clause 1(h), dealing with the promotion of national policies, should be elevated to first place. Other suggested revisions were:

- A commitment to sustainability in line with the guidelines and outputs from UNCED;
- Redefining the balance of objectives to give greater emphasis to value-added processing and better conditions for trade through improved market intelligence and market information; and
- A greater recognition of the role of local communities and indigenous people on the one hand and, on the other, strengthened cooperation with NGOs in conservation, the timber trade and industry.

Various views were expressed about the Year 2000 Objective. Some believed that it should be given a pre-eminent place as the overall setting for the objectives of ITTO. Others felt that this would tend to place a limit of time on the objectives of ITTO which should, instead, be applicable well beyond 2000. Yet others wished it to be accompanied by a statement of the resources required to meet such a target.

The first session of the Preparatory Committee revealed the extent of the underlying differences among the participants. Appropriately diplomatic statements were made by the spokespersons for the producers and consumers, but many of the other speakers were not so restrained. Brazil preferred to see the Agreement retain its character as a commodity agreement. Ecuador wished the Agreement to provide mechanisms to ensure that the development objectives would be achieved; the 2000 target 'should be seen not as an end in itself but as a process and a means to provide financial resources for tropical countries through ITTO; there should be an equitable geographical spread of projects'.[5] Sweden urged that the successor Agreement should continue to be a commodity agreement; as such, ITTA should focus on the production chain from the forest to the end user. ITTO's Target 2000 still lacked substance; the strategy needed to be fleshed out.

Switzerland was even more outspoken. It had resisted calls for unilateral restrictions on tropical timber imports because it considered that they would be counter-productive; this attitude might change, however, if ITTO efforts to achieve Target 2000 failed. The criteria for financing projects should be tightened. The criteria should integrate trade in tropical timber and Target 2000; they should have a multiplier effect and should demonstrate that the rational use of tropical forest is attainable; and the results of projects must be widely disseminated and used for capacity building. The central objective of a new agreement should be sustainable management of tropical forest resources for export and the expansion of trade. The Guidelines for the Sustainable Management of Natural Tropical Forests should be made mandatory; equally it should be mandatory for members to submit national reports on how they fulfil their obligations under the Agreement. Switzerland was not convinced that ITTO was the institution or this the time to take up coverage of temperate timbers. It believed 'that the forest and environmental policies of the industrial countries and the universal validity of the principle of shared environmental costs would be better taken up during a negotiation for a global forest convention'.

The representative of the timber trade and industry indicated that there was now firm ground for consensus among the trade. There was global concern for tropical deforestation, its negative impact on the international timber trade, foreign exchange earnings and the socio-economic development of producing countries. ITTO remained 'the one organization with a unique role in promoting sustainable management of the forest resource base in order to enable international timber trade to be maintained at least at its present value and possibly with prospects of increased revenue'. The Trade disagreed with proposals to extend the scope of the ITTA to cover all forests. It was vital, instead, to consolidate the achievements of the Organization and build on them.

The environmental NGOs were also outspoken. The rate of tropical deforestation had increased since 1983. NGOs had found serious deficiencies in the implementation of ITTA, 1983. Although there had been useful progress in defining international standards, there had not been much change in the policies of either producers or consumers, nor had there been significant improvement in the practices of logging enterprises.

The NGOs considered that there was a need for a fundamental change in the scope and mandate of the ITTA in the following ways:

- The principal role should be to develop equitable trading agreements based on genuinely remunerative prices for all timber products. Other major objectives should relate to the monitoring of the transfer of technology and increased processing in tropical countries.
- All types of timber, including temperate and boreal, should be incorporated into the Agreement.
- The balance between project and policy work should be redressed: projects should be strictly limited to activities which would promote the development of the Organization's main objectives.
- Environmental and social issues should be addressed by national and international policies.
- ITTO should concentrate on those activities in which it had the greatest comparative advantage. It should reform itself to be effective in fostering 'fairness of trade and the maximization of benefits from all the world's forests'.

The issues were clearly still wide open. A number of important questions were unresolved: the proposal to expand the scope of the Agreement to cover the trade in all timber – tropical, temperate and boreal; the degree of prominence to be given to the Year 2000 Objective; the mention of finance and technology in relation to the Objective; whether and where the format and frequency of reporting should be specified; and the balance between ITTO's activities.

In two Working Papers, the Chairman summarized the possible scope of renegotiation and options for it. 'The following proposals for the scope of the new Agreement had been made: (a) No change. Tropical timber only, using 1983 definitions. (b) Tropical timber as now defined plus tropical conifers but excluding pulp. (c) Tropical timber plus tropical conifers plus non-timber forest products but excluding rubber, cocoa and other plantation/tree crops. (d) All timber – tropical, temperate and boreal.'

If members decided to expand the scope of the new Agreement, there were five possible courses of action: (a) to renew ITTA, 1983 and keep talking in ITTC; (b) to renegotiate ITTA, 1983 and keep talking in ITTC; (c) to renew or renegotiate but refer wider issues elsewhere (to FAO, UN: UNCTAD); (d) to renegotiate bringing all the issues into the scope of the debate; or (e) to request the UN General Assembly to establish an Intergovernmental Negotiating Committee for a Framework Convention on Forests with a protocol on trade issues.

The Second PrepCom started where the first had ended. After extensive discussions in the producer and consumer groups, two alternative texts were prepared and presented to the Committee as a parallel tabulation (including also the text of ITTA, 1983). These dramatically highlighted the differences between the two groups which are reflected in carefully argued statements made by the two spokesmen to the Committee, Mr Amha bin Buang of Malaysia and Mr Milton Drucker of the US.[6]

Mr Amha introduced his statement by emphasizing that the producers were of the view that the Year 2000 Objective should not be inscribed in the Agreement. 'Producers do not wish that this guiding and indicative target be turned into a deadline or ultimatum which would be made legally binding to members through its formal inscription in the new Agreement.' But, at the same time, he made it clear that their position on the inscription of the target in the Agreement did not in any way affect their existing commitment to the target. They considered that the existing provisions for reporting were adequate. 'Producers are clearly not in favour of any shift towards any structure of mandatory reporting that will undoubtedly impinge on the sovereignty of members.' They were of the view 'that there [had] been a clear overemphasis, if not an obsession, on forestry management and conservation under ITTA, 1983'; there were other areas of work which were as important – if not more so – to the promotion and achievement of sustainable management. These included the crucial aspects of marketing; pricing; market transparency; utilization of wood, particularly from lesser-known species; processing and value-added product development; transfer of technology; human resource development; and R&D. The producers considered:

> that it [was] extremely important that these substantive areas be given
> due attention under the successor Agreement so that an overall balance
> in the work of [the] Organization [would] be achieved that would
> enhance its usefulness and effectiveness ... the producers [were]
> committed to making the ITTA a truly international commodity
> agreement.

The producers believed that the ITTA should remain a commodity agreement, that it should focus on timber but that it should cover all timber. They had given 'deep and thorough thought to this position and have arrived at it on the basis of sound and compelling considerations'.

> As we are all aware, the current Agreement focuses exclusively on tropical timber. In appropriate circumstances, exclusiveness is a positive and desired state of affairs but in the case of tropical timber, the exclusiveness given is anything but positive.[7] Tropical timber has been given exclusive scrutiny by consumer members and the NGO community. In the light of the renegotiation of the current Agreement, there have been further moves to stifle the position of tropical timber with elaborate requisites of sustainability as well as new and additional conditionalities. *Indeed tropical timber and forests have been singled out and made a convenient scapegoat for all the environmental woes facing our planet.* All that we, producers, have got out of the exclusiveness are largely criticisms, allegations, pressures, conditionalities, and threats of boycotts, bans and restrictions on tropical timber. For as long as ITTA is kept in its existing form, tropical timber and forests will continue to be caged in with conditionalities of sustainability and linkages to the environment, human rights and democracy, and be used as a convenient beating boy for the environmental problems of the world.[8]
>
> Producers can no longer tolerate this state of affairs, as it is not a true and fair reflection of the actual environmental crisis facing the globe. We can no longer accept that the sustainability of temperate and other timbers and forest could be a matter of convenient assumption. *Given the revealing report by the WWF and other relevant studies of the problems facing temperate and other forests, we are convinced that* the state of affairs of other timbers should be the subject of the same vigorous scrutiny and conditionalities. We are sick and tired of the application of double standards; one for tropical timber and none for temperate and other timbers, and we want this eliminated under the Successor Agreement...
>
> The current Agreement is glaringly inconsistent with all the watershed decisions reached at UNCED... To ensure the continuing relevance of ITTA in the years ahead, it is therefore imperative that its scope be expanded to include all types of timbers.
>
> Among the most serious concerns which producers have on the current Agreement is its serious implications on trade in tropical timber... Given the conditionalities being clinically applied to tropical timber, the marketing of the product will be made less favourable to that of temperate timber and like products which are not currently subject to the same scrutiny and conditionalities, thereby distorting the conditions of competition between these products. This discriminatory trend is clearly inconsistent with the provisions of GATT and will aggravate the declining share of tropical timber in the international timber market.

The consumers advanced contrary arguments on the most significant points. Mr Drucker began by listing problems and then addressed the solutions proposed in the consumers' draft.

First, the problems. Tropical timber was facing a difficult marketing problem as the environment movement gathered steam in developed countries. This was a problem that both groups must face together.

> We understand that tropical timber has an important role to play in the world economy; it is an important source of income for producing countries; it is an important resource for consuming countries; a renewable resource of great beauty and structural value. But, if the consuming public is afraid to purchase tropical timber or tropical timber products because they fear that they are destroying an important environmental resource, then the work we are doing here – no matter how good – will be for naught. What we do here must persuade the world that it is not merely safe to purchase tropical timber, but helpful.

Mr Drucker went on to discuss the difficulties of attaining market transparency for tropical timber because of the small scale and dispersed nature of the industry. 'Market transparency will be helpful to policy planners and governments in both developed and developing countries. Despite our common commitment to market transparency in the ITTA, 1983, we have not yet achieved full transparency in tropical timber markets. This is a problem we must continue to deal with.'

> This leads me to the closely related problem of reporting on tropical timber trade and related activities, which has been seen by producers as an imposition on producer countries. This is an unfortunate viewpoint. Full and accurate reporting will contribute immeasurably to the development of plans and programmes to achieve sustainability. Without a thorough knowledge of what is happening in the countries represented in the ITTO, it will be difficult to provide assistance to them in dealing with the problems they have in achieving sustainability. The lack of data and lack of resources to acquire data in producer countries is well known in the ITTO. Now we must do something about it – together.

Projects were vital to the Organization but it was difficult to raise money for them. Funding could be maintained 'if, and only if, (a) we can show [that] the projects meet the objectives that all ITTA members have signed up for, in other words the objectives that are in the agreement itself; and (b) [that] the projects are developed and administered in a manner which meets the highest standards. Failing either of those necessary conditions, we are doomed to a lack of funds.'

Trying to fix 'fair' prices had been a spectacular failure in the International Tin Council. 'No international organization nor any government can raise timber prices in the world market. Those prices will be set by supply and demand.'

Next, the solutions. Demonstrating progress towards the Year 2000 Objective would be 'vital in maintaining credibility of the organization over the next several years and establishing it as an important force in the international community in future decades'.

> We have already seen consumers in large numbers turn away from tropical timber and parliaments feel the need to seek legislative remedies to the issue. If the ITTO can provide anything, it can provide an alternative to national legislation which may turn out to be inefficient, even painful, and which will not improve the ability of nations to develop solutions co-operatively together. We urge producers to continue to work with us as they have since Bali when producers supported the Year 2000 Target with us... It would indeed be a step back, a large step back, if we failed to find a means in the ITTO II of doing what we set out together to do.

The consumers had produced some suggested text on these three issues – market transparency, reporting and the Year 2000 Objective – and a framework within which the Council could act to improve the project cycle.

The last issue was the suggestion that the ITTA should be converted into an agreement covering all types of timber. The consumer group continued to oppose this. Mr Drucker went on to explain their reasons:

> We are concerned that the producer group feels such a move would alleviate unfair discrimination. We cannot disagree more. First, if the perceived discrimination relates to national legislation ... I can only say that I doubt an all-timber agreement would have any impact on the national legislation which tropical timber exporters will face if we cannot demonstrate that a shift towards sustainability is a reality. Second, we do not agree that an agreement on all timber is possible without examining the multiple uses of forest – in other words a global forest convention. Third, we strongly support a global debate on forest issues ... but such a discussion would have a very different venue from this. Fourth, to the extent that discrimination within ITTO is an issue, we need to come to grips with it in the normal course of our work, to listen to one another, and to work to resolve the perception of fact of unfair discrimination. Changing the nature of the ITTA may in fact cause discrimination as smaller producers lose their voice in an organization dominated by the largest temperate, boreal, and tropical timber producers. Fifth, ITTO assistance is focussed on those most in need. A larger membership would focus on other issues and other countries. Sixth, the ITTA, 1983 is completely consistent with the Forest Principles we worked so hard on in UNCED. Nothing in the Forest Principles text is inconsistent with agreements which cover regions, types of forests, or types of trade.

Mr Drucker concluded as follows:

> The Consumers Group has converted the ITTA, 1983 into an ITTA II of which we will all be proud. We imagine our Producer colleagues have been engaged in a similar effort. We look forward to working with them to make our joint effort one which governments, trade and public interest groups will look to in the future as a model of co-operation in the areas of trade and environment.

In the Negotiating Conference itself, the negotiations were protracted and difficult. Progress was reported to both the 14th and 15th Sessions of the Council in May and November 1993. At its 14th Session, the Council had before it the report of the First Part of the Negotiating Conference. The Chairman, Mr Samuel Kwasi Appiah of Ghana, urged that the new agreement needed to be 'founded in mutual consideration and trust, and not in polarized views and political postures. Negotiations could only succeed if the vigilance of one nation over its interests was moderated by the recognition that the interests of others must also be served.'[9] The Council debated the report received from the First Part and established a 'reconciliation group' which prepared a draft to be transmitted to the Second Part in June.

Mr Drucker directed attention to the three main issues of contention. The consumers still did not believe that it would be fruitful to include all forests in the successor agreement, or to apply Target 2000 to these forests, but nevertheless were prepared to reconsider the sustainability of their own forest practices and to commit themselves accordingly. On the question of trade discrimination, the consumers were prepared to ensure that nothing in the new agreement would allow such discrimination, and that no internationally traded timber should face discrimination on grounds of origin. On resources, they would endeavour to make some unearmarked funds available to the Council, provided that the decision-making process was transparent and the funds were well administered.

This did not satisfy the spokesman for the NGOs (Mr Grant Bryan Rosoman), who found that the statement of the consumers lacked substance. It was vague about sustainable management; it proposed neither targets nor timetables, and the question of double standards was not addressed. He considered that the link between trade and sustainable management was well substantiated. Accordingly, the successor agreement must apply to all forests and timber; it should include provisions for better market transparency and information sharing, further processing and value-added production, transfer of technology, improved market access for products from sustainably managed sources, full internalization of social and environmental costs, and support for developing country members.

The spokesman for the trade associations, Mr Alle H Stoit, gave his full support for Target 2000. There was no future for the trade without sustainable management and, without demand, there was no incentive for sustainable management. He agreed with the conclusions of the London Environmental Economics Centre (LEEC) report[10] – trade was not a major cause of forest destruction. Fundamental conditions were 'firstly access to markets, secondly market improvement for value-added products and lesser-used species and, thirdly, harmony between the ambitions of producers and the needs of consumers'. He added that, 'at present the consumers were being effectively manipulated by the anti-tropical hardwood campaigns of the environmental NGOs', and he called upon the NGOs to use their knowledge in a positive way.

The 15th Session of the Council followed the Third Part of the Negotiating Committee. It was addressed by the President of the UN Conference, HE Mr Wisber Loeis of Indonesia, who gave an account of the course of events since the 14th Session. He recalled that:

> at the outset of the negotiations, pessimism was intense, to the point where some nations were threatening not to adhere to a new Agreement if it did not embrace their cherished principles. Others spoke of liquidation of the whole Organization. He had actively sought to discourage such lines of thought. ITTO was the one organization whose mandate was truly imbued with conservation-oriented imperatives. As such, it did not deserve to have its future potential damaged by a gloomy lack of interest in one of the world's most treasured resources. The moral and material resources of the international community had been invested in ITTO with the full expectation of making full progress against some of the ecological problems now besetting mankind.[11]

With his President's Informal Working Group he had endeavoured to 'galvanize a small consultative group to exert leverage on the larger body of Delegations. This was his effort to stem the tide of divergent views on the issues of scope, trade discrimination and financial resources before these reached the point of no return'. During the Third Part, this group was enlarged and negotiated on the basis of the 'President's Revised Discussion Paper' and many others specially prepared on difficult issues – the Year 2000 Objective, trade discrimination, and information sharing. By this time, significant progress had been made on the establishment of the 'Year 2000 Objective Fund'. Mr Loeis admitted that the negotiations were still battling with the *scope* of the new agreement. His revised discussion paper presented the following compromise proposals:

- Limitation of the scope of a new agreement to tropical timber and tropical forests.
- Focus on the Year 2000 Objective and sustainable development thereafter.
- Progress through *rational* use and trade in the resource to ensure its conservation.
- Recognition that, in consonance with UNCED Principles, increased funds should be forthcoming from the Organization to assist achievement of the sustainable management objective.
- UNCED Principles also required a *global* perspective to the foregoing objectives, involving tropical and non-tropical forest owners alike.

- ITTO should provide a universal forum for discussion of trading issues relating to all forests and for sharing of information.

Mr Loeis observed that mere recognition of a need was inadequate. Practical measures were essential. For this reason, the President had proposed a minimal expansion, or what was more justifiably termed an *enhancement* of the agreement. 'Such a forum would prove sincerity and realism in the new agreement by addressing the major issues; but, at present, compromise was not possible if some persisted in rejecting even such a minimal enhancement and others desired a radical expansion.'

Statements from the NGOs were more pessimistic and hinted at their probable loss of interest in ITTO. Mr William E Mankin of the Global Forest Policy Project was blunt:

> The future of ITTO was clouded with the last opportunity to give it a new lease of life about to vanish... A decade of genesis and growth had still left ITTO in search of an identity, despite constructive criticism from various quarters... Only one member country seemed willing to support the Organization fully... Commitments to sustainability had produced few measurable results and ITTO and its members appeared to have been mere spectators to world debate and action on important trade-related issues.

Mr Mankin portrayed an atmosphere of declining interest in ITTO's mission and little progress on the disputed core issue of the negotiations. 'NGOs had worked hard to convince members that the negotiations offered an opportunity to chart a new course. Yet it was hard to tell whether NGOs' textual proposals had been seriously considered or not.'

> The scope should be broadened to include all types of timber because, if this was not the case, action focusing on tropical timber alone could be fully justified. Indeed, such discriminatory actions would proliferate, and members would never have grounds to appeal to GATT. If concerned citizens did not see progress towards sustainable management of tropical forests at a speed they desired, naturally they would promote actions to hasten the process.
>
> Consumer nations with temperate timber resources could not escape such trade related moves by seeking to keep the scope of the successor agreement narrow. Already legislative programmes and certification schemes included all types of timber. The ITTO was at a crossroads; adaptation to change and innovation would ensure its survival, but failure to do so would see the world moving on without it.

In the end, good sense and compromise prevailed. The ITTA, 1994 was concluded at the Fourth Part of the UN Conference for the Negotiation of a Successor Agreement to the ITTA, 1983.[12]

BOX 9.1 THE OBJECTIVES OF THE ITTA, 1994

Recognizing the sovereignty of members over their natural resources, as defined in Principle 1 (a) of the Non-Legally Binding Authoritative Statement of Principles for a Global Consensus on the Management, Conservation and Sustainable Development of all Types of Forests, the objectives of the International Tropical Timber Agreement, 1994 (hereinafter referred to as 'this Agreement') are:

(a) To provide an effective framework for consultation, international cooperation and policy development among all members with regard to all relevant aspects of the world timber economy;

(b) To provide a forum for consultation to promote non-discriminatory timber trade practices;

(c) To contribute to the process of sustainable development;

(d) To enhance the capacity of members to implement a strategy for achieving exports of tropical timber and timber products from sustainably managed sources by the year 2000;

(e) To promote the expansion and diversification of international trade in tropical timber from sustainable sources by improving the structural conditions in international markets, by taking into account, on the one hand, a long term increase in consumption and continuity of supplies, and, on the other, prices which reflect the costs of sustainable forest management and which are remunerative and equitable for members, and the improvement of market access;

(f) To promote and support research and development with a view to improving forest management and efficiency of wood utilization as well as increasing the capacity to conserve and enhance other forest values in timber producing tropical forests;

(g) To develop and contribute towards mechanisms for the provision of new and additional financial resources and expertise needed to enhance the capacity of producing members to attain the objectives of this Agreement;

(h) To improve market intelligence with a view to ensuring greater transparency in the international timber market, including the gathering, compilation, and dissemination of trade related data, including data related to species being traded;

(i) To promote increased and further processing of tropical timber from sustainable sources in producing member countries with a view to promoting their industrialization and thereby increasing their employment opportunities and export earnings;

(j) To encourage members to support and develop industrial tropical timber reforestation and forest management activities as well as rehabilitation of degraded forest land, with due regard for the interests of local communities dependent on forest resources;

(k) To improve marketing and distribution of tropical timber exports from sustainably managed sources;

(l) To encourage members to develop national policies aimed at sustainable utilization and conservation of timber producing forests and their genetic resources and at maintaining the ecological balance in the regions concerned, in the context of tropical timber trade;

(m) To promote the access to, and transfer of, technologies and technical cooperation to implement the objectives of this Agreement, including on concessional and preferential terms and conditions, as mutually agreed; and

(n) To encourage information-sharing on the international timber market.

Source: ITTA, 1994, Article 1.

Between them, the Preamble and Objectives (Box 9.1) deal with most of the contentious issues. The Agreement is confined to tropical forests but there are many points which deal with the timber trade as a whole. The Year 2000 Objective is mentioned in the Preamble but, in the Objectives, it only occurs in Article 1(d) in the context of enhancing capacity. Most of the important words and phrases occur in these two parts of the Agreement (a strategy for achieving international trade in tropical timber from sustainably managed sources; comparable and appropriate guidelines and criteria; global perspective in order to improve transparency in the international timber market; commitment of all members to achieve exports of tropical timber products from sustainable managed sources by the year 2000; new and additional financial resources; strengthen the framework of international cooperation; forum for consultation to promote non-discriminatory timber trade practices; improving the structural conditions in international markets; prices which reflect the costs of sustainable management; greater transparency; and increased and further processing).

A most important development was Article 21 establishing the Bali Partnership Fund. Clause 4 gave guidance about how it should be spent:

> In allocating resources of the Fund, Council shall take into account:
> a. The special needs of members whose forest sector's contribution to their economies is adversely affected by the implementation of the strategy for achieving exports of tropical timber and timber products from sustainably managed sources by the year 2000;
> b. The needs of members with significant forest areas who establish conservation programmes in timber producing countries.

At last some substance had been given to the original idea of a reforestation fund.

Another very significant advance was made when a Formal Statement was prepared by the consumer members which was appended to the Agreement. In this they affirm:

- All States listed below commit to implement appropriate guidelines and criteria for sustainable management of their forests comparable to those developed by the International Tropical Timber Organization;
- Those States which have already achieved a high standard of sustainable management of their forest commit to maintain and enhance the sustainable management of their forests;
- Other States commit to the national objective of achieving sustainable management of their forests by the Year 2000; and
- Appropriate resources should be provided to developing countries to enable them to achieve the objective of sustainable forest management.

Australia, Austria, Canada, China, European Community, Belgium/ Luxembourg, Denmark, France, Germany, Greece, Ireland, Italy, Netherlands, Portugal, Spain, United Kingdom of Great Britain and

Northern Ireland, Finland, Japan, New Zealand, Norway, Republic of Korea, Russian Federation, Sweden, Switzerland, United States of America.

A comparable statement was made on the sustainable management of forests in the European Union by the European Union.

The Chairman of the two PrepComs, Mr Andrew Bennett, had this to say:[13]

> ITTO has been a very effective forum. At times when it has been difficult to talk rationally about forest issues elsewhere, it's always been possible in ITTO to meet with and talk to colleagues from a wide range of backgrounds.
>
> In many respects, there's not a lot of difference between the two agreements, even though we spent a great deal of effort in redrafting... When we work with the new Agreement, it will be interesting to see whether the activities of ITTO are noticeably different.
>
> I think the renegotiation process opened some rifts and created some tensions in the Organization. The last session of the renegotiation was particularly bruising and left a lot of people feeling unhappy. We must rebuild some bridges; we've got to strengthen the consensus and learn to trust each other again.

10

Has the tropical timber trade any leverage? Policies 1991–1995

'Sustainable forest management is the process of managing permanent forest land to achieve one or more clearly specified objectives of management with regard to the production of a continuous flow of desired forest products and services without undue reduction of its inherent values and future productivity and without undue undesirable effects on the physical and social environment'
Definition of sustainable forest management adopted by the
International Tropical Timber Council

While the negotiations for the ITTA, 1994 were proceeding, ITTO was far from inactive. The period between 1987 and 1990 had seen many initiatives and a new wave began in 1991 during the Quito session. This time saw the birth of the Year 2000 Objective and four new and important activities: the development of ITTO's pioneering set of criteria and indicators, the study on 'Economic Linkages between the International Trade in Tropical Timber and the Sustainable Management of Tropical Forests' carried out by LEEC, the Forest Resource Assessment study and the examination of incentives in Ghana.[1] These started before UNCED and, with the exception of the criteria and indicators, were not completed until after the conclusion of the ITTA, 1994. The gradual maturing of these policy documents provided a backcloth for many of the discussions in Council between the Quito session until after the consideration of the Mid-Term Review in 1995.

Criteria and Indicators

Among the elements of Decision 3(X) reached in Quito was the item which led to the first criteria and indicators for tropical forests or, indeed, for any type of forest. It read as follows:

> Consider, at its XI Session, the conclusions and recommendations of an expert panel convened by the Executive Director to develop methods of defining and measuring sustainable tropical forest management and decide on any further initiatives to ensure the attainment of the Year 2000 Target.

Accordingly, the Executive Director commissioned a consultants' report.[2]

The report of the consultants began by giving a discussion of the concept of sustainability. This, in retrospect, covers most of the issues which later became prominent in the international discussions about criteria and indicators.

> The concept of sustainable management is not simple. This is due partly to the different interpretations put upon the terms used and partly to the complications inherent in the idea. In order, therefore, to make the arguments as clear as possible, we have adopted the following definitions of the three operative words:
>
> * *Sustainable:* Condition which maintains the level of some identified feature (quantity, quality or activity) into the indefinite future.
> * *Forest:* Land naturally stocked or planted with trees which provide environmental services, conserve biological diversity and produce timber and non-timber forest products. The term is taken to include natural forests and plantations but to exclude scattered trees on farmland.[3]
> * *Management:* Process which follows a deliberate course of action in pursuit of a specific goal.
>
> 'Sustainable Management' implies that the management system itself should be sustainable. The system should, therefore, be sufficiently flexible to cope with the unexpected, with future events and with changing knowledge and techniques.
>
> The issue becomes particularly complicated because the concept has been applied to different purposes of management, to different scales of operation and to different time frames. Some preliminary discussion of these points will help in assessing the possible criteria that may be chosen to evaluate progress towards the sustainable management of tropical forests. But, in spite of every precaution, there are always likely to be differences of opinion between various interest groups about the range of values that should be included in any consideration of sustainability.
>
> (a) The time frame. It is not possible to say, at any single moment, that a particular management system will be sustained. *Sustainable management is a process, which can only be assessed over a period of time.* Management procedures, which are satisfactory today, may break down tomorrow. Those that are unsatisfactory today may be rapidly improved. In fact, efficient and sustainable management may involve a continuous process of change if it is to be sustained.
>
> (b) Use and management. Sustainable use is not the same as sustainable management. The present use of some areas of forest may be sustainable, but without effective management there is no assurance that it will be sustained.
>
> (c) Objectives of management. There are a number of possible *features* which it may be desirable to maintain in any area of forest: the

full species complement of the forest (biological diversity); the forest structure; its biological productivity; its quality as the living space of forest-dwelling people; the potential to produce identified products; soil fertility; and climatic, physical and hydrological regulatory functions. The sustainability of each of these is governed by different criteria.

(d) Single- or multi-purpose management. Any tract of forest may be managed principally to maintain one of these features or for different combinations of them (single- or multi-purpose management). Adverse effects can be minimized but some trade-off is often unavoidable.

(e) Land zonation. In terms of ultimate sustainability, there is a relationship between the pattern of land use in a country and the objectives of management of individual tracts of forest. The loss of certain qualities necessitated by single-purpose management can be offset by appropriate allocation of land for the maintenance of these qualities. This applies particularly to the preservation of biodiversity.

(f) National or forest level. It follows that the term sustainable forest management can be validly used at two levels: to refer to a particular tract of forest or to all the forests in a country. There should, therefore, be criteria to assess 'sustainable forest management' at two levels: for the nation and for the individual tract of forest (management unit).

- At national level the criteria should be designed to make it possible to assess the effectiveness of land allocation in achieving and maintaining an optimum mix of the qualities identified above.
- At the level of the forest tract they should be designed to make it possible to assess the effectiveness of forest management in maintaining the potential of the forest to discharge the function chosen for it. This implies different criteria for different primary objectives of management.

It follows, therefore, that:

- Sustainable forest management depends upon the clear definition of the objectives of management; the firm implementation of these objectives; regular monitoring of performance; and periodical adjustment of procedures so that the result conforms to the objectives.
- If the criteria to assess sustainability are to be complete, they should cover the national pattern of land use; forest management for different primary uses; and forest management for various multiple uses.

The consultants next attempted a concise definition of sustainable forest management, which was adopted, slightly modified, by the Expert Panel and, thereafter, by the Council, as follows:

Sustainable forest management is the process of managing permanent forest land to achieve one or more clearly specified objectives of management with regard to the production of a continuous flow of desired forest products and services without any undue reduction of its inherent values and future productivity and without undue undesirable effects on the physical and social environment.[4]

They added the following qualification:

It should be stressed that there can be no absolute interpretation of this, or perhaps any other, definition of sustainable forest management. It is likely to be interpreted differently by different interest groups. But, more important, attaining the ideal is a process of continuous adjustment.

The consultants then went on to discuss the assessment of sustainable forest management. It depended upon: elements of national policy (political commitment, economic policies, policies for land use and forestry, and supporting legislation); broad public support for these policies; an appropriate government structure to ensure the effective implementation of sectoral policies, and the necessary connections and interactions between sectors; prescriptions for action based on 'best practice'; guidelines for all important operations in both natural and planted forests; and a high standard of implementation. This in turn depended upon adequate capacity in terms of institutions, staff numbers, qualifications, commitment and conditions of service, and a high standard of implementation. (The need for finance to support this capacity was, of course, implied.) The success or failure of sustainable forest management could only be assessed in the field by the examination of chosen indicators. They emphasized that no complete assessment could be made at any single time. 'Attaining sustainability [was] a process of progressive refinement of the quality of both prescriptions and implementation, a refinement that [would be] brought about by a conscientious examination of the indicators of performance – by monitoring and feedback. There [would] always be an element of risk due to changing economic conditions (markets, currencies, etc) and political circumstances (changes of government, new policies, etc).'

The consultants thought the ITTO Guidelines for the Sustainable Management of Tropical Forests provided a thorough but general review of preconditions for sustainable forest management; they specified many highly desirable elements, which would contribute to it; and they provided a framework of principles within which measurable criteria could be identified. The material from the guidelines was useful in providing a checklist of the items that should be part of effective reporting procedures.[5]

The second part of the consultants' terms of reference required them 'to decide on any further initiatives to ensure the attainment of the Year 2000 Target'. Under this heading, they introduced the matter of reporting and wrote as follows:

If progress is to be made towards attaining sustainability by the year 2000, it is important that tropical forest countries should keep under regular review the status of their own forest lands and of the management of their forests, and that this information should be made available to ITTO at regular intervals. In order to make a proper diagnosis of the situation, both tropical countries and ITTO require three categories of information.[6] These categories are:

- Category 1: basic information on policies, legislation and all statutory standards;
- Category 2: regular information on any changes in (1), about prescriptions for action and about action taken; and
- Category 3: detailed and regular information on the changes in a number of identified *indicators of sustainability*.

In addition, it would be valuable if the reports were to include *specific targets* for future performance and *details of progress* in attaining these targets.

The Report went further. It recommended 'that all the Producer States which were members of ITTO should provide the Secretariat and the Council with all the documentation necessary to demonstrate their progress towards sustainable forest management and towards Target 2000 and listed the material that should be provided under each Category. The basic documentation (Category 1) should be provided once only; that in Categories 2 and 3 annually'. The checklist of topics to be covered under Category 3 is reproduced in Box 10.1.

The final section of the report discussed some implications and made three recommendations:

We have given a definition of sustainable forest management which depends upon a zonation of forest uses, the establishment of TPAs and a permanent production forest, the precise determination of objectives of management and the application of high environmental standards to all operations. If all these conditions are met, it should be possible to manage the whole forest estate of any country in a manner that is sustainable and, equally, to manage any part of it in a way which meets the specific objectives of management in a sustainable manner. Our particular task is to identify measurable indicators by which success or failure can be measured.

An initial difficulty is that some may not accept our definition and that there is bound to be disagreement about the relative importance to be given to the various values of tropical forest and to the kinds of trade-off that are permissible. This is a matter that cannot be resolved; and ITTO must reach its own judgement of the exact balance to be given to these various factors.

The sets of criteria we have identified should, however, help it to do so. The documentation proposed under Category 1 should give the basic information about the adequacy of policies, legislation and prescriptions. That under Category 2 should provide an annual update on this and statistics which will enable an evaluation to be made of the extent to which these various prescriptions are applied in practice. It is Category 3

Box 10.1 PROPOSED CHECKLIST FOR REPORT ON PROGRESS TOWARDS SUSTAINABLE MANAGEMENT

1   Country
2   Constitutional, policy and legal provisions
3   Land resources and uses (categories)
4   Forest resources
    4.1 Establishment of TPA
    4.2 Establishment of Production Forest
    4.3 Other categories and uses
5   Forestry policies, strategies and programmes
6   Forestry in the national economy
    6.1 Contribution to GDP
    6.2 Revenue and export earnings
    6.3 Employment and income
7   National forest inventory and resource monitoring
8   Management of natural forests for production
    8.1 Objectives of management
    8.2 Management and operational plans
    8.3 Management and operational inventories
    8.4 Determination of AAC and prescribed cut
    8.5 Harvesting and production
    8.6 Reforestation and rehabilitation
    8.7 Boundary demarcation and protection
9   Management of planted forests
    9.1 Objectives of establishment and management
    9.2 Age class distribution
    9.3 Management and operational plans
    9.4 Determination of AAC and prescribed cut
    9.5 Harvesting and production
    9.6 Silvicultural programme
    9.7 Forest protection

10 Management of forests for biological diversity
    10.1 Environmental management
    10.2 Environmental stability and quality
    10.3 Soil properties
    10.4 Hydrological characteristics
    10.5 Microclimatic changes
11 Social matters
    11.1 Objectives
    11.2 Consultation
    11.3 Adaptive measures
12 Research and development
    12.1 Biological productivity and production
    12.2 Growth and yield/growth modelling
    12.3 Inventory projection/yield prediction
    12.4 Effects of harvesting and silviculture on biodiversity
    12.5 Integrated experimental studies
    12.6 Forest operations
    12.7 Economic and policy analysis
    12.8 Multiple use management
    12.9 Forest protection
13 Forest administration
    13.1 Organisation and staffing
    13.2 Human resources development
14 Financial aspects
    14.1 Budget and performance
    14.2 Development fund and programme
    14.3 Investment incentives for agricultural and forestry development
15 Conclusion

Source: Mok and Poore, 1991.

that is designed to give quantifiable measures by which success can be judged. It is these, therefore, that will be the ultimate criteria and arbiters of sustainable forest management; nothing less precise will give the indicators that are required.

Here lies the problem. There is very little valid information about many of the variables that must be measured. Few tropical countries have the capacity either in money or human resources to provide all the information required in Category 2, far less in Category 3; and even world-wide there may not be sufficient skills, for example in plant and animal taxonomy. Yet, if the presence of sustainable forest management is to be successfully diagnosed, these critical indicators must be measured. It is evident that a compromise must be reached: in the immediate future there must be the very selective identification of indicators which can be measured at once, relatively easily and cheaply, but which give a sufficiently accurate indication of sustainability.

We have gone as far as we can in making suggestions; ultimately many of the variables are site specific. But we need to know the principles by which they should be selected.

Often, recommendations for more research are an excuse for delay and procrastination. At one and the same time there is a need to make the best use of such information as we possess and to set in train the programmes to obtain the information we really need.

We recommend that ITTO should take action on three fronts:

- To set up a reporting procedure based on the information that can be obtained now that will give the best possible measure of sustainable forest management based upon all the elements in Criteria 1 and 2;
- To establish in the different continents integrated research programmes to examine in detail the main variables in models of national sustainable forest management in order to identify exactly what variables should be measured to provide the most accurate and economical measure of progress towards sustainable forest management.
- Use its influence to ensure that people with the necessary skills are available as soon as possible to implement all these essentials.

The consultants' report was submitted in August, 1991 and was considered by an Expert Panel in the Hague in September; the Panel's report, with the consultants' report as an annex, was submitted to the 11th Session of the ITTC in November–December, 1991. The Expert Panel trod more delicately than the consultants.

In developing the criteria presented in this report, the Panel made a number of basic assumptions about how the criteria could be used in practice. One of these assumptions was that the criteria would provide a framework for Member countries to voluntarily submit reports on their efforts and achievements in promoting and attaining sustainable tropical forest management and trade in tropical timber from sustainably managed resources in the context of Target 2000. In this, the panel was mindful that to be effective, reporting on sustainability would have to take place in the context of the co-operation and transparency provided by the ITTO forum. In particular, it noted that the text of the Decision invited member countries to submit reports on their progress. It was felt that ITTO could create the environment where it was in the interest of all relevant parties to respond to requests for reports noting that the open forum of the ITTC itself provides for the necessary consideration of the reports of the individual members.

The Panel selected 'five broad criteria' at the national level: the forest resource base; the continuity of the flow of forest products; the level of environmental control; socio-economic effects; and institutional frameworks. At the level of the management unit there were also five criteria: resource security; the continuity of timber production; the conservation of biological diversity; an acceptable level of environmental impact; and positive socio-economic benefits. The Panel's report, however, failed to be definite about indicators and contented itself with a list of 'possible indicators'.

In its conclusions, the Panel 'was acutely aware that even the reporting envisaged would place considerable strains on the human and technical resources of many of its members. Consequently, the panel felt that its suggested reporting format should be treated as a tentative rather than a definitive one that would need assessment by Council.'

It continued:

> The adoption of Target 2000 means that ITTO and its members have at most only 8 years to achieve sustainable management in the tropical forests. Given this timetable, the panel felt that ITTO would not be able to afford the luxury of developing after several years a practicable reporting system through experience. Thus, the panel thought that there was considerable merit in both the producer and consumer members of ITTO joining together in an accelerated experiment in the development of the reporting framework and its subsequent application in setting priorities in accelerating progress towards target 2000… Such action would also promote a framework of co-operation that guaranteed both national sovereignty and equity in the attainment of Target 2000. Tropical timbers compete with timbers from other geographic regions in many markets and equity demands that tropical producers be not saddled with costs not likely to be encountered by their competitors. More important, however, is the opportunity such an experiment would provide … to share experiences and rapidly develop more relevant and effective frameworks for achieving sustainable forest management.

The Panel's report ends with the odd twist: 'Such experiences would, of course, be highly relevant to consumer countries as they struggle with the issues associated with sustainable development.'

Thereafter, the proposals became embroiled in the bureaucracy of Council Decisions and might, indeed, have remained exclusive to the members of ITTO had not the Executive Director taken the initiative of publishing the definition, the criteria and the examples of indicators as No. 3 of the ITTO Policy and Development Series. After the Expert Panel, the course of events was as follows.

At the 11th Session, the ITTC considered the report of the Expert Panel, requested that the Executive Director should transmit the definition and criteria for sustainability for further refinement, and that a Consultative Panel should be established during the 12th Session to 'examine the issues relevant to sustainable management'.

There does not seem to be any record of the meeting of the Consultative Panel called during the 12th Session, but it appears from the relevant Decision[7] that some members, both producer and consumer, were not yet prepared to commit themselves to the proposed criteria and indicators, far less to agree to regular reporting as recommended by the consultants. These members expressed the view 'that more utility would be gained by refining the text of the criteria on the basis of operational experience and field application'. This was the beginning of a continuing tug of war between those who wished to apply the existing criteria and indicators at once, however imperfect they might be, in order to gain some appreciation of progress, and those who used the provisional nature of the indicators as a reason for not reporting. In spite of these objections, however, the Council decided on several positive steps: to adopt the definition of sustainable forest management; to disseminate widely the criteria and indicators publication; and to invite members to apply them in the field and exchange experience. Consumer members were invited to make funds available.[8]

This decision *did* launch the criteria and indicators, however tentatively, as a tool to be used in the 1995 reports on progress. The Council, however, did not take the opportunity to accept the 'format' for reporting provided for them in the consultants' report and referred this matter to the next meeting of Council. The reporting format which emerged was not helpful in that it in no way emphasized those items which would give evidence of progress towards the Year 2000 Objective in the field (see Box 10.2).[9] There are profound differences in both content and emphasis between this and the recommendations in Box 10.1. The recommended format resulted, as we shall see, in reports for 1995 that were little use in assessing progress; and the mistake was compounded in 2000 by recommending the same format for reporting in that year. This was the unfortunate consequence of the ITTO weakness of making political decisions on what are essentially technical issues.

The later development of the 'criteria and indicators movement', both outside and inside ITTO is described in Chapter 11.

BOX 10.2 REPORTING FORMAT FOR ANNUAL REPORTS

Reporting Format

The following reporting format has been provided by the Council to assist Members in the preparation of annual reports as provided for under Articles 27 and 28 of the Agreement and for informing Council of progress towards meeting the Year 2000 Target.

In commending this format, Council noted that Members would only be expected to report once annually on those aspects relevant to their own specific circumstances and which are within the scope of the Agreement. Council also noted that the availability of data is presently limited for some Members and that this would affect their reporting capability.

Council also noted that having submitted their first reports utilizing this format, members would only be expected to provide updated information in subsequent annual reports.

Reporting Format

1 Introduction/Summary
2 Institutional and policy framework
   • Legal and institutional framework for national tropical forest policy and implementation of tropical forest management plans.
   • Relationship of ITTO Decisions, Target 2000 and ITTO Guidelines to national forest policy.
   • Relevant legislation and other measures affecting trade in tropical timber.
   • Measures to increase the efficiency of tropical timber utilisation and promote production of value-added timber products.
3 Tropical forest Resource Base
   • Areas and distribution of protection forests, production forests and plantations and their relation to national goals and targets.
   • Plantation establishment targets and annual planting regimes.
4 Production and trade of logs, sawnwood, veneer, plywood, fibreboard
   • Production level by major products by species groups and estimates of future production.
   • Export and import values and volumes.
   • Prices for major products by species groups.
   • Stocks.
   • Share of tropical timber in total timber trade.
   • Annual trends in timber production from the forest, consumption and international trade.
5 International cooperation
   • International financial and technical co-operation relevant to tropical forest management and international trade in tropical timber.
   • Research and development in tropical forest resource issues.
   • Measures to increase production and utilization efficiency including measures to increase value-added in producer countries.
6 Environmental measures
   • Environmental legislation and policies as related to tropical timber; environmental assessments, regulations for forest operations, and other relevant measures.
7 Socio-economic effects
   • Economic flows associated with production and/or use of tropical timber.
   • Provisions for involvement of local communities.
   • General economic conditions which affect supply and demand of wood products.

Source: Annex to ITTC Decision 5(XIII).

The Development of Forest Resource Accounting

It had become evident that if the Year 2000 Objective were to become a reality it would be necessary to develop some kind of reporting mechanism and a framework of criteria upon which progress might be judged. Two lines were pursued: (a) a system of 'Forest Resource Accounting: monitoring forest condition and management' which never became fully operational, for it was largely overtaken by (b) the more internationally fashionable criteria and indicators for sustainable forest management – which, equally, could be used as a basis for reporting. The origins of the latter have been described above.

The system for Forest Resource Accounting (FRA)[10] was developed and tested by the UK Overseas Development Administration (ODA) in collaboration with IIED and the World Conservation Monitoring Centre (WCMC). An ITTO study on the system started in October 1991, the main objective being to develop a forest resource accounting methodology for carrying out accurate and quantitative assessments of the condition of a country's forest resource, enabling tropical timber producer countries to carry out a comprehensive evaluation of the standard of management of their forest resources and to update such an evaluation on a regular basis.

The final report of the study was not presented to the Council until its 15th Session in November 1993 and it was not published until 1994.[11] The Committee for Economic Information and Market Intelligence 'reviewed and took note of the progress report [sic]. The Committee expressed satisfaction with the report.' However, neither the report nor its recommendations went any further. It seems that, by this time, the international movement towards defining (and even using) criteria and indicators had gained such momentum that it was effectively unstoppable. There was also a marked reluctance to report – a reluctance that persists to this day.[12]

Economic Linkages

The proposal that there should be a study on the economic linkages between the international trade in tropical timber and the sustainable management of tropical forests was first proposed at the 10th Session in Bali and was later adopted as an ITTO activity. The study was carried out by LEEC in London and was presented to the Council in November 1992.

Just as foresters saw the Sarawak Mission as defining what should be ITTO's attitude towards forest management, so did the trade regard the LEEC study on economic linkages as defining the reality of the limited influence of the international timber trade on forest survival. Some in the trade also believed that the study would make the environmental community more aware that it was not enough to target the international timber trade alone.

The objectives of the study were: to establish in detail the international trade flows in tropical timber; to discern the underlying policies and market conditions that determine these flows; to assess future trends in the tropical timber trade; to analyse the nature and extent of the impact of the current trade in tropical timber on the management of tropical forests; to evaluate whether trade policy interventions were effective instruments for directly altering the incentives for sustainable management of tropical forests; to evaluate new trade policy measures designed for funding resource transfers in support of sustainable management initiatives in producer countries; and to examine the institutional aspects of these policy interventions and measures, particularly their consistency and feasibility in light of established regimes of international cooperation in international trade.

Summarized very briefly, the main findings were as follows.[13]

The international trade in tropical timber was not a major cause of deforestation; the conversion of forests to other uses such as agriculture was much more significant. The volume of tropical timber traded internationally was small and declining, an increasing proportion being used domestically. Only 11 per cent of logs and 12 per cent of sawnwood was exported.[14]

As resources were depleted and domestic consumption increased, the price of logs could be expected to rise and the amount exported to fall. But the prices of tropical hardwood products would probably be depressed by the increasing amounts of temperate timber reaching the market and, perhaps, by other wood and non-wood substitutes. Consequently, the processors of tropical logs would find it difficult to make a profit or to expand their export markets.

There were, however, real grounds for concern about the future of the forest. Among them were: continuing excessive exploitation and rapid depletion of tropical production forests in many regions; the effect of bad harvesting practices on non-timber forest values and in creating incentives to convert land to other uses; and a failure to match the expansion in processing capacity to the economic availability of timber stocks. Economic, forestry and land use policies in producer countries had distorted the incentives for sustainable management and had failed to curb the wider environmental problems associated with timber extraction. Trade policy distortions in producer and consumer countries had exacerbated the situation.

The key factor in reversing these trends was to ensure that there were proper economic *incentives* in favour of the efficient and sustainable management of production forests – to maximize the long-term potential of the forest to generate income, and to internalize significant environmental costs. Greater returns from trade in tropical timber and its products would encourage sustainable management in production forests, but only if policies for sustainable management were adopted and implemented. In contrast, trade interventions such as bans, taxes and quantitative limitations actually *restricted* trade, reduced incentives for sustainable timber management and might indeed *increase* overall deforestation.

The report was cautious about the effectiveness of trade policy measures in reducing deforestation: the most direct effect of trade measures was on cross-border flows and prices, and changes in these would have little influence on deforestation. There was, however, a place for *trade-related incentives* for sustainable management. These would be most effective if (a) they complemented domestic policies and regulations for sustainable forest

BOX 10.3 LEEC: REASONS FOR COUNTRY CERTIFICATION

Reasons why country certification may be more effective and workable:

1   Certification at the country level will be less costly and more easy to implement compared to other certification schemes. Periodic inspection tours by an internationally certified team, monitoring at custom ports, and review of forest policy and management plans would probably be sufficient for such a scheme to be effective.

2   Producer countries would find country certification schemes more politically acceptable, provided that:
   (a)   under international auspices producer countries could help determine the certification scheme as well as any verification process;
   (b)   the certificate of origin could then be issued by the exporting country, or companies authorized by that country;
   (c)   as it is effectively a *national* sustainable management plan that is being certified, it would be up to the producer country to address adequately the problems posed by production from conversion forests, plantations, reafforestation and so forth, but once this plan is internationally verified, *all* timber products from all types of forests in the country would be certified;
   (d)   it would be up to the producer country to ensure compliance with the sustainable management plan, and, in cooperation with 'independent inspection', the relevant forest authorities would have primary responsibility for monitoring operations at the concession and industry level; and
   (e)   in exchange for adopting the scheme and being certified, an exporting country would hope to receive improved access to international markets for their 'sustainably managed' products and, hopefully, international financial assistance to implement their sustainable management plans. This would provide an incentive for producer countries to adopt this policy approach.

3   Consumer countries may also find country certification schemes more feasible to implement as:
   (a)   under international auspices, such as ITTO, consumer countries could help determine the certification scheme as well as any verification process;
   (b)   individual timber trade products would not have to be certified – all trade products from a certified country could be safely imported, and any inspection could be conducted as routine port of entry (ie customs) procedures;
   (c)   consumer countries would be assured that country certification would require a policy commitment by producer country governments to manage their production forests sustainably under the ITTO guidelines and Target 2000, viable national plans to implement this policy, and a mandate to correct domestic market and policy failures that encourage timber-related deforestation; and
   (d)   it would be easier to target bilateral and multilateral assistance for sustainable forest management – ie such flows could now be conditional on producer countries complying with the certification scheme.

In sum, certification ought to be used as a means to facilitate trade in sustainably produced tropical timber – not as a means to restrict trade. Product labelling and concession certification are too cumbersome and restrictive to assist trade-related incentives of sustainable management, and may also fail to get adequate producer country participation. On the other hand, country certification would require the active participation and verification of producer countries. In exchange, producer countries could qualify for additional financial assistance to implement sustainable management plans, as well as be assured of better market access for their exports.

Source: LEEC, 1992.

management; (b) they improved access to import markets to ensure maximum value-added for sustainably produced exported products; and (c) they helped producer countries to obtain additional financial resources to implement national plans for the sustainable management of their production forests.

The authors of the report commended to ITTO a number of measures that would reinforce the commitment made to the Year 2000 Objective:

- ITTO should encourage the establishment of a *country certification scheme*. (The arguments in favour of country certification are reproduced in Box 10.3).[15] The main objective would be to verify that a producer country is *implementing policies, regulations and management plans that ensure substantial progress towards the Year 2000 Target*. In return for satisfying these conditions *all tropical timber products* of that country would be certified as 'sustainably' produced.
- As part of the review of their policies to enable certification, producer countries should be required to examine: (a) the implications of their existing domestic forestry policies and regulations on timber-related deforestation; and (b) the extent to which their existing forest sector trade policies may also be affecting deforestation. In order to show progress towards achieving the Year 2000 Objective, producer countries ought to *demonstrate a commitment to correcting these policy distortions*.
- Producer countries that qualify for country certification *should be allowed better access for these products in the import markets of consumer countries*, eg the removal of tariff and non-tariff barriers. Consumer countries could also promote the use of tropical timber imports from certified countries through information and market intelligence campaigns.
- Certified producer countries should also qualify for the *additional financial assistance* they require for implementing national sustainable management plans and policies. (Quoted estimates range from US$300 million to 1300 million annually.)

The report went on to recommend that *ITTO should take the lead in promoting additional international financial assistance for those countries that demonstrate progress towards the Year 2000 Target and are duly certified*.

There was a full and fertile discussion of the report in the course of the Annual Market Discussion. It was recommended that it and associated papers should be circulated to all members and that the subject should be raised at the next meeting of Council, but it did not in fact appear on the agenda of that meeting. It is probable that the arguments in this report have had some influence on the general attitudes of members, but its apparent slide from the attention of the Council is a sad example of the way in which valuable recommendations may be lost in the complicated deliberations of the ITTC. (For details on the fate of LEEC's proposals on certification, see Appendix 2).

11

The many roads from UNCED

'The Intergovernmental Forum on Forests, the successor to the Intergovernmental Panel on Forests (IPF), was established in 1997… [It] reworked much of the ground covered by the IPF – which constitutes, in fact, the whole range of problems confronting world forests!'

International Developments: 1992–1995

Starting with the Earth Summit, there was a prodigious amount of international activity focusing on the world's forests. It is well beyond the scope of this book to give a blow-by-blow account of all the meetings and initiatives that took place between 1992 and 2000; this has been admirably done in the publications of the Commonwealth Forestry Association – *The World Forests: Initiatives since Rio (Rio+2; Rio+5; Rio+8).*[1] But an outline will be given of the main events that affected the evolution of the idea of sustainable forest management or that were of particular concern to ITTO.

Global Partnerships on Forests

After Rio, many countries were concerned about the highly polarized positions taken by the developing and the industrialized countries on the forest issue and felt the need to build trust and confidence among various countries on their perspectives on forests. Against this background, the Government of Indonesia took the bold initiative in 1993 of organizing a landmark international conference on 'Global Partnerships on Forests' in order to review major issues and to build bridges between the North and the South.

Ministerial Conference on the Protection of Forests in Europe

An early initiative was taken by European States which held, in June 1993, a Ministerial Conference on the Protection of Forests in Europe, generally known as the Helsinki Conference, or more recently as MCPFE.[2] This resulted in a General Declaration which referred extensively to the Forest Principles and Agenda 21 and faithfully reflected the sentiments in these two documents.

But the Declaration was almost entirely concerned with the management of European forests and did not mention the possible effects of the policies of European countries on forests outside Europe. There is no reference to international trade, nor any direct reference to aid. Nevertheless, it is notable that this declaration predated and anticipated the formal statement made by the European Union in the final stages of the negotiation of ITTA, 1994 (see Chapter 9). Moreover, the signatories declared their intention to 'participate in, and promote, international activities towards a global convention on the management, conservation and sustainable development of all types of forest'.

The conference also passed four resolutions, the first on General Guidelines for the Sustainable Management of Forests in Europe, the second on General Guidelines for the Conservation of the Biodiversity of European Forests, the third on Forestry Cooperation with Countries with Economies in Transition, and the last on Strategies for a Process of Long-term Adaptation of Forests in Europe to Climate Change. All four formally committed the countries concerned to some general principles and specific actions. The guidelines on sustainable forest management included a definition:

> For the purposes of this resolution, 'sustainable management' means the stewardship and use of forests and forest lands in a way, and at a rate, that maintains their biodiversity, productivity, regeneration capacity, vitality and their potential to fulfil, now and in the future, relevant ecological, economic and social functions, at local, national, and global levels, and that does not cause damage to other ecosystems.[3]

On sustainable management, the commitment was towards preparing national or regional guidelines, and 'common measures consistent with these guidelines that would favour the production, use and marketing of products from forest under sustainable management'. As policies developed further, this commitment was realized by defining and applying criteria and indicators. This was the stimulus for the 'Helsinki Process', which produced the first set of criteria and indicators to be developed outside ITTO.

The CSCE Conference, Montreal

Another significant event was an initiative of the Government of Canada which had, undoubtedly, been stung by criticisms of forestry in British Columbia. In October 1993, Natural Resources Canada hosted a conference in Montreal under the auspices of the Conference on Security and Cooperation in Europe (CSCE) which was devoted to discussing the sustainable development of boreal and temperate forests.[4] A very important objective was to generate criteria and indicators for these forests. The input from ITTO was considerable. Representatives from seven ITTO producer nations attended, as well as members of the Secretariat, and the present author gave one of the two invited papers on criteria for the sustainable development of forests.[5] Two workshops were held – one on social and economic and the other on environmental criteria/indicators, resulting in two sets of suggested criteria and indicators which were to have profound and lasting effects as the foundation of the 'Montreal Process'. ITTO can therefore, with justification, claim to have been an important stimulus to and originator of the international move towards the adoption of criteria and indicators for sustainable forest management.

The original quest of many people at that time was to find a few simple quantitative indicators, which, like temperature and blood pressure in human beings, would define the 'health' of the forest. This was an elusive quest. In the social and economic fields, it proved very difficult to identify any simple quantitative indicators. The environmental field was different. It was generally possible to identify quantitative indicators for most variables but it would often have required long and expensive local research programmes to attach reliable figures to such indicators. The report of the environmental workshop reads like a programme for research for the whole of the succeeding century!

This was the start of the long (and still incomplete) task of identifying useful indicators which can reasonably be applied in practice. From the beginning, the Helsinki and ITTO Processes took paths that diverged from that of the Montreal Process. Helsinki and ITTO, relying as they both did on formal governmental agreements, sought pragmatic schemes that could be applied immediately in the field, whereas the Montreal process looked at 'desirable' indicators – ones that would give firm quantitative measures eventually, even though they were not necessarily practicable at the time. It is, nevertheless, quite remarkable how far the results of the various processes have converged, so that they are, for practical purposes, identical in their essentials. In Chapter 13 we shall return to later developments in criteria and indicators and discuss some of the salient features of those that are now being used.

The Commission on Sustainable Development

The Commission on Sustainable Development (CSD) was set up under the aegis of the Economic and Social Council of the United Nations (ECOSOC) by Resolution 47/191 of the UN General Assembly, with the task of overseeing the implementation of the various outputs from UNCED. The CSD established the IPF (Intergovernmental Panel on Forests, 1995–1997), followed by the IFF (Intergovernmental Forum on Forests, 1997–2000) which was then superseded by the United Nations Forum on Forests (UNFF). An outline of the most recent activities is given at the end of this chapter.

The India–United Kingdom Initiative

There were two important intergovernmental actions in preparation for the meeting of CSD in 1995: the India–United Kingdom Initiative towards Sustainable Forestry and the Malaysia–Canada Initiative – the Intergovernmental Working Group on Forests. Both were efforts to heal the rift between North and South, which had characterized international forestry discussions. In fact, the innumerable forestry meetings that took place between

1992 and 1995 were very important in developing personal friendships and a feeling of trust between participants – significant agents in improving international understanding.

An output of the India–UK Initiative was a resolution adopting a 'Framework for National Reporting to the 3rd Session of the CSD'. It was recommended that each country should give a short description of its most important and relevant achievements from among the topics in the Forest Principles and Agenda 21 – relevant quantitative or qualitative markers of progress should be inserted where available.

The Malaysia–Canada Initiative

The Malaysia–Canada Initiative (the Intergovernmental Working Group on Forests) identified possible 'options' – action was apparently too strong a word! – for the CSD in seven separate fields, which together encapsulated all the important issues in the prevailing forestry debate. They were: (a) forest conservation, enhancing forest cover and the role of forests in meeting basic human needs; (b) criteria and indicators for sustainable forest management; (c) trade and environment; (d) approaches to mobilizing financial resources and technology transfer; (e) institutional linkages; (f) participation and transparency in forest management; and (g) comprehensive cross-sectoral integration including land use planning and management and the influence of policies external to the forest sector. The output of this initiative largely set the agenda for the IPF.

The first three of these options were of particular relevance to ITTO. That on trade was perhaps the most innovative; the details are given in Box 11.1.

Center for International Forestry Research

A very significant development was the setting up of the Center for International Forestry Research (CIFOR) as part of the Consultative Group on International Agricultural Research (CGIAR) system. At last, forestry was to be given a research centre to match in status the several institutes that had made such a contribution to the advancement of agriculture. CIFOR held a policy dialogue on science, forests and sustainable development in December 1994 to identify and consider key research needs for the support of sustainable forest management to meet human needs now and in the future.[6] Since its establishment it has developed very significant multi-disciplinary research programmes throughout the tropics, some in collaboration with ITTO.

World Commission on Sustainable Development

At the same time, an independent body of eminent personalities was established – the World Commission on Sustainable Development (WCSD) – which, although working closely with the IPF, was able to consider issues that were deemed too delicate to be handled in formal international meetings.

BOX 11.1 MALAYSIA–CANADA INITIATIVE:

OPTIONS FOR TRADE

TRADE AND THE MANAGEMENT, CONSERVATION AND SUSTAINABLE DEVELOPMENT OF ALL TYPES OF FORESTS

Options

1   An appropriate body or bodies within the UN system of other multilateral organizations, with the participation of other stakeholders, could examine trade and other measures that affect countries' abilities to attain the management, conservation and sustainable development of all types of forests.
2   The incorporation of the full costs of the management, conservation and sustainable development of all types of forests into market mechanisms and prices may facilitate making trade and environmental policies mutually reinforcing and thus support more effectively the management ... [of] forests. Such incorporation of costs has been the subject of research by a range of private, governmental, and international organizations. The CSD could encourage additional practical research in this area by an appropriate body or bodies within the UN system or other multilateral organizations.
3   An appropriate body ... could undertake an examination of policies related to subsidies, taxes, tariffs and related mechanisms in the forest and forest-related sectors, with a view to avoiding discrimination and in order to help ensure that countries' abilities to attain the management ... [of] forests is not impaired.
4   An appropriate international body could examine, with the participation of governments, NGOs and the private forestry industry sector, issues related to the potential harmonisation and the use of certification programmes as a means of promoting the management ... [of] forest. In this context, the timing, cost-effectiveness and market, social, environmental and economic impacts of various certification approaches should also be considered, having due regard to the different circumstances of countries, forest types and management practices.
5   An appropriate body ... could examine the issue of illegal trade in forest products, plant and animal species and genetic resources, with a view to suggesting means of combating such trade.

Source: Grayson, 1993.

Food and Agriculture Organization

As the UN agency with particular responsibility for forestry, FAO continued after UNCED to play its traditional roles and was appointed as Task Manager for Chapter 11 of Agenda 21. The changing shape of the TFAP has been described in Chapter 8. By 1995, the name was gradually changing to NFAP (National Forests Action Programme) as it came to include temperate and boreal as well as tropical forests.[7] The National Forest Programme (NFP) concept attracted strong support in international debates. A survey by FAO in 1999 found that while many, indeed most, countries stated that they had an NFP, the implementation in developing countries was very weak.[8]

Surely, by 1995, there should have been sufficient material to move immediately into action? But, as we shall see, the talking continued.

Forest Stewardship Council

A development of rather a different kind was the establishment of the Forest Stewardship Council (FSC).[9] It came into being to provide consumers with reliable information about forest products and their sources in response to public concern about forest degradation and the confusing proliferation of labelling schemes. It, like the labelling schemes, was based on the assumption that the quality of forest management would be significantly improved by the market pressure generated by consumer preferences. Its founders believed that it would have a more rapid and profound effect on forest management than intergovernmental agreements and meetings.

In early 1990 there was a move, largely powered by WWF, to establish an accreditation agency for timber from sustainably managed sources,[10] and the name Forest Stewardship Council was proposed for it. After consultation, the FSC was established in Toronto in October 1993 under an international board that represented many forestry interests. Unfortunately, both in their consultation and in membership of the board, the timber trade was neglected – an error of judgement that harmed the credibility of the new organization. The objective of the FSC was, and is, to promote good forest management throughout the world. It does this by evaluating, accrediting and monitoring certification organizations, which, in their turn, inspect forest operations and certify that forest products have come from well-managed forests. The FSC does not itself certify forest management or products; its mandate is to set a code of practice for certification, to accredit the certifiers, and to promote the development of national standards for forest management for the purposes of certification.

Starting in 1991, the FSC has developed and refined a set of principles and criteria that are intended to apply to all tropical, temperate and boreal forests. Local factors are taken into account in the evaluation of individual forests; national standards cover more detailed provisions and interpretation. The FSC and its accredited certifiers do not insist on perfection in satisfying the principles and criteria, but major failures in individual principles will normally disqualify a candidate from certification. The principles are designed to ensure that forests are managed in ways that are environmentally appropriate, socially beneficial and economically viable. They cover: compliance with national and international laws and FSC principles; respect for tenure and use rights and the rights of indigenous peoples; the enhancement of the well-being of forest workers and local communities; efficient use of the products and services of the forest to ensure economic viability and environmental and social benefits; the conservation of biological diversity, water, soils and special features; maintenance of the ecological functions and integrity of the forest; formulation and implementation of management plans; monitoring and assessment; and the maintenance of natural forests. A tenth was added later on plantations. The principles and criteria cover broadly the same ground as the various sets of criteria and indicators.

The FSC mark is widely considered to give a reliable indication that wood products come from sustainably managed sources, and a number of national certification schemes have allied themselves with the FSC principles and criteria. But the FSC has, from the beginning, been rather jealous and protective of its position. It effectively stifled an interesting initiative of the Tropical Forest Foundation in Washington to establish an award for *progress towards* sustainable forest management in the tropics, on the grounds that this would confuse and undermine its efforts to certify forests that were actually sustainably managed.

International Developments: after 1995

Forestry saw (or suffered from) a prodigious spate of international dialogue during this period.[11] The IPF had been given the mandate to deal with the following matters:

1   Implementation of UNCED decisions relating to forests at the national and international levels, including an examination of sectoral and cross-sectoral linkages.
2   International cooperation in financial assistance and technology transfer.
3   Scientific research, forest assessment, and development of criteria and indicators for sustainable forest management.
4   Trade and environment in relation to forest products and services.
5   International organizations and multilateral institutions, and instruments, including appropriate legal mechanisms.

It delivered its final report to the CSD in March 1997,[12] comprising a discussion and proposals for action under each of these heads.

Under heading 1, there were five subjects (number of proposals for action in brackets): (a) progress through national forest and land uses programmes (9); (b) underlying causes of deforestation (16); traditional forest-related knowledge (18); fragile ecosystems affected by desertification and drought (7); impact of airborne pollution on forests (5); and needs and requirements of developing and other countries with low forest cover (11). One suggestion – that a 'diagnostic framework' should be used to identify the causes of deforestation in particular local circumstances – has been adapted by ITTO (see below) to identify the factors which are critical in hindering the progress of a country towards sustainable forest management.

Under heading 3 were a number of items of interest to ITTO, for example the assessment and valuation of multiple benefits, and criteria and indicators. It is noteworthy that the issue of a 'core' set of criteria and indicators remained unresolved in this gathering.

Item 4, dealing with trade, contained some observations of considerable relevance to ITTO:

> The Panel discussed the following options for action relating to a
> possible agreement for forest products from all types of forests, based

on non-discriminatory rules and multilaterally agreed procedures, without reaching a consensus on these or other possible procedures:

(i) To take note of the ITTA of 1994, in particular the commitment made by ITTO members to review the scope of the agreement four years after its entry into force on 1 January 1997.

(ii) To explore the possibility of extending the concept of the Year 2000 Objective of ITTA for all types of forests.

(iii) To explore the possibility of an international agreement on trade in forest products from all types of forests.

(iv) To examine the possibility of further initiatives in trade liberalization within the auspices of the World Trade Organization.

(v) To explore, within an intergovernmental forum on forests, intergovernmental negotiating committee or other arrangements decided upon at an appropriate time, the possibilities of promoting the management, conservation and sustainable development of all types of forests and trade in forest products in the context of an international, comprehensive and legally binding instrument on all types of forests.

The IFF, the successor to the IPF, was established in 1997. Its mandate was:

> To promote and facilitate the implementation of the proposals for action of the IPF; review, monitor and report on progress in the management, conservation and sustainable development of all types of forests; consider matters left pending as regards the programme elements of the IPF; identify the possible elements of and work towards consensus on international arrangements and mechanisms, for example, a legally binding instrument.

The IFF reworked much of the ground covered by the IPF – which constitutes, in fact, the whole range of problems confronting world forests![13] Topics included: decision making; policy tools such as national forest programmes and criteria and indicators; information and public participation; scientific knowledge; traditional forest-related knowledge; monitoring, assessing and reporting on progress towards sustainable forest management; forest resources and their management; deforestation and forest degradation; forest health and productivity; rehabilitation and maintenance of forest cover; and forest conservation and the protection of representative and unique types of forests. Another set of issues related to international cooperation and capacity building, financial resources, international trade and the transfer of environmentally sound technologies. Perhaps most important is the question of future international mechanisms. A large volume of material was assembled on these subjects and there were many supporting meetings to discuss particular issues. For example, in San Jose, Costa Rica on underlying causes of deforestation; in Canberra on forest conservation and protected areas; and in San Juan, Puerto Rico on protected areas. There were also significant meetings to discuss implementation – in particular, the Six-Country Initiative on Implementation of the IPF Proposals for Action (1998, Baden-Baden, Germany).

At the final session of the IFF, in February 2000, consensus was reached on many of these issues; others were left pending – the legal and financial mechanisms towards sustainable forest management proved too controversial for final resolution; others 'need[ed] further clarification'. The items referring to ITTO in the report of the IPF are not repeated in the report of IFF.

To quote from Rio+8:

> Despite a lack of agreement on legal and financial mechanisms [for] forests, one of the most important legacies of the IFF process is the political will to reach consensus on many complex issues concerning international forest policy. Perhaps the most concrete outcome ... is the wide-ranging set of proposals for action... These proposals [about 130 in number] provide governments, international organizations, private sector and all other major groups [with] guidance on how to further develop, implement and coordinate national and international policies on sustainable forest management.

Particularly interesting, in view of the debate about a possible new 'forest instrument', were the conclusions in the last section about institutional arrangements. At great length, it made the point that the best way forward was to strengthen the capacity of existing organizations to work in cooperative and complementary partnerships, each using its comparative advantage. In particular, the work of the Inter-Agency Task Force was commended. Sanity had prevailed!

The report then went on to propose the setting up of a further successor body 'which may be called the United Nations Forum on Forests'.[14]

The UNFF was accordingly set up on 18 October 2000, by resolution 2000/35 of ECOSOC, as a subsidiary body of ECOSOC 'with particular emphasis on the *implementation* of the proposals for action of the IPF/IFF'. The Collaborative Partnership on Forests (CPF) was built on the foundations of the Inter-Agency Task Force. To assist in the work of the UNFF, a meeting was held in Bonn, Germany in December 2000 – the Eight-Country Initiative on Shaping the Programme of Work of the United Nations Forum on Forests. At its first meeting, in June 2001, it set up a phased programme to address its work.

The outcome of these meetings must influence the future work of ITTO and will almost certainly affect the form of any future agreement when the ITTA, 1994 is renegotiated. But ITTO will have a reciprocal influence through its membership of the CPF.

Meanwhile, while the world has talked, the existing organizations continued with practical implementation.

FAO has played an important part in the IPF/IFF/UNFF processes, recently as focal agency within the Collaborative Partnership on Forests. The FAO Strategic Plan for Forestry was approved in 1999 and has formed the basis for its Medium-Term Plan 2002–07. It aims: to continue, with clear priorities, to execute FAO's mandated roles viz global information on food and agriculture, a neutral forum for policy and technical dialogue, and policy advice and technical assistance; and to build mutually beneficial partnerships with other organizations. Priority areas are: the Global Forest Resources Assessment; statistics on production, trade and capacity for wood products; criteria and indicators for forest management; certification; the strengthening of institutions; and national forest programmes. In all of these, there has been close collaboration with ITTO, for example in aspects of the planning of FRA 2000, in preparation of the Tropical Timber Bulletin, and in criteria and indicators, certification, reduced impact logging, and National Forest Programmes.

ITTO was also active, as we shall find in the following chapters.

ITTO's road to 1995: ENGOs diverge

---

*'WWF had supported a widening of the ambit of the successor agreement to
the ITTA, 1983 from the outset of the negotiations and the parallel extensions
of the type of products within the mandate… Negotiators had decided not to
follow the WWF recommendations and this marked a parting of the ways'*
Gordon Shepherd, WWF

This chapter examines ITTO's work from late 1993 until 1995, the date fixed
for the Mid-Term Review of progress towards the Year 2000 Objective. During
this period, some of the principal issues to concern the Council were trade
discrimination, a move towards using CITES to regulate trade in certain species,
the cost of attaining sustainable management, the disaffection of the
international NGOs with progress in ITTO, and the Mid-Term Review itself.

The 15th Session of Council – Yokohama, November 1993

The 15th Session fell between Part 3 and Part 4 of the Negotiating Conference.
Council was informed that the Helsinki Process was under way and that the
CSD had been established. A working party on certification was established –
not before time.

Trade discrimination

Once again, there were sharp interchanges on the question of possible trade
discrimination. In the course of giving his country report, the representative of
the Netherlands asked for views on his government's policy document, first
introduced in 1991. The response took him by surprise. It was a trenchant
condemnation of the Netherlands Framework Agreement on Tropical Timber
(NFATT) – an important element in his country's policy. The spokesman of the
producers, 'registered [the producers'] strongest concern over and rejection of
the framework agreement'.[1] His statement revealed the producers' continuing,
intense distrust of the motives of the consumers in introducing any measures
connected with labelling or certification.

> [Producers] found it amazing that the Netherlands, a country with very little residual natural forest, had seen fit to formulate a policy governing the trade in products from a type of forest which did not exist in her own country... The original policy document of 1991 ... was inconsistent with the Forestry Principles... The producers believed the policy should be immediately replaced with one that was pragmatic, fair and treated all types of forest impartially. In [the producers'] view the NFATT was an unfair, premature and counter-productive measure which would injure the producers' legitimate interest in the marketability of the products of their forests.

He listed the bases of their concern:

> It was a restrictive trade measure against tropical timber, setting a deadline of 31.12.95 beyond which tropical timber from countries not considered to manage their forests sustainably would be refused access to the Netherlands. Producers believed that, to the extent that they would be aggrieved by this unilateral action, they would be entitled to take countervailing actions under the provisions of GATT.
>
> It deliberately discriminated against tropical timber vis à vis temperate timber and would inevitably cause the prices of the former to fall relative to the latter. This was a problem of double standards of precisely the kind the producers were trying to overcome in the course of the negotiations for a successor agreement. At the very least this initiative could have been made applicable to all types of timber and timber products.

The spokesman believed that there was a conflict between ITTO's Year 2000 Objective, to which the Netherlands was a party, and the NFATT's target of 1995; the two targets were irreconcilable.

His criticism was also directed at 'the other signatories of the NFATT, in particular WWF and IUCN'. He remarked that the producers had been encouraged by the support given by the NGOs for the elimination of double standards; he urged them, accordingly, to be consistent in approaching the issue of the NFATT. This statement was widely supported by other producer nations and by the African Timber Organization (ATO). There is no record of any reply by the Netherlands or the NGOs.

The 16th Session of Council – Cartagena, May 1994

The withdrawal of WWF

The next Session – the 16th, held in May 1994 – was the first to take place after the conclusion of ITTA, 1994. It witnessed the final break between WWF and the Council. Since the very beginning, WWF had faithfully attended Council sessions, sometimes fielding a delegation of formidable strength.[2] It fell to Mr Gordon Shepherd to make the announcement. He informed the Council:

> that WWF had been involved with the Organization since before its inception and had tried to mobilize staff resources to help it fulfil its mandate. However, recently, the cost had reached about SwFr100,000 per annum. WWF had supported a widening of the ambit of the successor agreement to the ITTA, 1983 from the outset of the negotiations and the parallel extensions of the type of products within the mandate. It saw such a change as an improvement in the capacity of the Organization to perform effectively, also as creating a true partnership.
>
> Negotiators had decided not to follow the WWF recommendations and this marked a parting of the ways. Therefore, the WWF would be reducing its participation in ITTO activities and considering re-allocating resources to better options such as the Biodiversity Convention and CITES. WWF found it difficult to see how the new ITTA, 1994 could make a useful contribution and proposed that the ITTO concentrate on trade-related work which would increase market transparency.

He referred to the Certification Working Party.

> WWF was convinced that consumers wished to buy timber products from sustainably managed forests. It had called in 1989 for the international timber trade to be based on sustainably managed forests by 1995... WWF believed that this should apply to all forests and it had helped co-sponsor the Forest Stewardship Council. Any certification scheme had to be equitable, voluntary, open, independent and publicly accountable; the FSC had been constituted with these principles in mind as a certifier of certifiers.[3]

Many felt that the complete withdrawal of WWF – and it was accompanied by other international NGOs – was an unfortunate act of pique at failing to get exactly what they wanted. It would have been more statesmanlike to retain some less expensive connection and to continue the valuable contribution they had made to the debate. The attempt to attain sustainable forest management by 2000 was likely to be difficult enough; the 1989 suggestion that it could be realized by 1995 had been pie in the sky.

There were no statements from environmental NGOs at the 17th and 18th Sessions.

Involvement with CITES

It was perhaps a coincidence that it was at the Cartagena Session that CITES made its first significant appearance on the Council agenda. But the way in which it did so was not helpful and provoked another extended and tendentious debate.

CITES was introduced by a report from the representative of the TFF, who had just returned from a meeting of the Plants Committee of CITES. He told the meeting that the following timber trees had been proposed to the Committee for CITES listing: 'ramin' (*Gonystylus bancanus*): proposed by the Netherlands, supported by the Philippines, opposed by Malaysia; *Khaya* and *Entandophragma* spp: proposed by Germany and possibly the US; *Swietenia macrophylla*: proposed by the US and possibly the Netherlands, supported by El Salvador and Colombia, opposed by Mexico; and *Dalbergia melanoxylon*: proposed by Kenya. Ramin (and probably the others) were proposed for Appendix II of the Convention – 'species whose survival is not yet threatened but may become so. Trade allowed subject to licensing.'

The spokesman for the producers voiced concern at the new proposals. CITES was a legally binding convention. Listing involved the monitoring of commercial transactions and control through elaborate measures including direct prohibition on trade. Effects were therefore far reaching. It was reasonable if proposals for listing came from states in which the species occurred, 'since these states acted voluntarily out of a genuine interest in the protection of these species within their frontiers. However, the recent trend towards proposals originating from non-range states was unhealthy, since range states felt pressurized, and suspicious that the proponents had ulterior motives unconnected with any concern for protection for the species whose listing was proposed.'[4] He considered that using CITES was 'tolerable' for species of wild fauna, although even with them it had not been particularly successful. He went on to talk about 'ramin' – an important timber species of the peat swamp forests in Sarawak.

> Listing did not greatly affect the interest of non-range states but could damage the economies of range states through its impact on foreign exchange earnings and employment. In the view of many producers, certain proponents were 'playing to the gallery' on green issues. But a rash listing could lead to a stigma on the species in question. Producers doubted the sincerity of such proposals and the 'ramin' (*Gonystylus bancanus*) was a good case in point...
>
> Mistrust was inevitable, especially as since 1992 [when the species was first proposed for listing] Malaysia had further strengthened sustainable management of her forests... Not only the conservation status of ramin but that of the whole Malaysian natural tropical forest had been markedly improved.[5]

Mr Sollo of Cameroon then intervened, referring to 'proposals for listing by consumer countries driven by pressure groups of species of *Khaya* and *Entandophragma*'.

> In addition, a list of redwood species including 'iroko' (*Chlorophora excelsa*) had circulated in Central African countries and in the headquarters of certain NGOs with the aim of using CITES to foster a

boycott campaign against African tropical timbers on the spurious pretext of extinction risk or biodiversity loss.

African members vehemently opposed methods of this kind that lacked any scientific basis. Fluctuations of wood volumes on the international market could just as easily be due to a switch to other species as to increasing scarcity or, indeed, to the deterioration of public statistical services recently afflicting many African countries.

In order to confirm any perceived risk of extinction of a species, it was necessary to possess inventory results showing low frequency of abundance in both upper and lower canopy layers... Did the listing proponents have such information? As far as Cameroon was concerned, inventories conducted in three-quarters of the productive forest had demonstrated that *Khaya* spp ... were regularly and abundantly represented in all the strata. Moreover, the distribution pattern conformed to the floristic diversity commonly found in this type of forest.

This was an unfortunate episode. It became clear from the discussion which followed that the debate had developed from an imperfect account of the proposals before CITES and, indeed, by incomplete understanding within the Council of the procedures followed by CITES and the nature of the three appendices.[6] But it had a constructive outcome. It resulted in a Decision[7] that led to a greater exchange of information about the species proposed for listing and to more consistent approaches being taken in CITES and ITTO towards any timber species thought to be endangered by trade. Collaboration was carried further in the next two Sessions of Council. At the 17th Session, in November 1994, it was decided[8] that relations with CITES should be much closer and that meetings should be scheduled to complement each other.

CITES was invited to make a presentation at the next session. This duly took place at the 18th Session in Accra in May 1995 when a member of the CITES Secretariat addressed the Council. He explained that there were 128 signatories, of which 50 were also members of ITTO. He stressed that it was important that country representatives should take the same position at meetings of the two organizations. He described the purpose of the convention and the nature of the three appendices. If a species were listed in Appendix I, international trade was prohibited except for non-commercial purposes. For species listed in Appendix II, commercial trade was permitted but controlled. An export or re-export certificate issued by the country from which the species was being exported or re-exported was required. Appendix II was in fact a 'sustainable utilization' appendix; it was the country of origin that must decide how much trade in the species should be permitted. Species could only be listed in Appendix III at the request of a country in which it occurred; commercial trade was permitted but controlled; listing was intended to enable other countries to assist the listing country in the protection of the species in question.

It was, however, possible for measures to be introduced unilaterally. He quoted the case of the US, which had introduced strict regulations on the import of wild birds that might prevent species being imported to the US even if they had been legally exported from their country of origin. The EU also had more severe restrictions on the importing of some species in Appendix II. 'This additional legislation may have led to the impression that Appendix I and Appendix II of the convention were the same and amounted to an import ban.'[9]

The original criteria for listing were adopted in 1977. These were vague, making the discussion of the inclusion of timber species in CITES more difficult than necessary. New criteria had been adopted in 1994 that would make decisions easier and less contentious, including those for the inclusion of timber species. At the last Conference of the Parties it was recognized that CITES might play a role in the trade of some timber species but that time should be taken to consider the implications. A Timber Working Group of 18 members had thus been established to look at issues related to CITES and the timber trade.

The presentation led to a long and constructive discussion and to the adoption of a decision[10] in favour of close cooperation between ITTO and the CITES Timber Working Group, on which ITTO should be represented.

The 19th Session of Council – Yokohama, November 1995

The discussions in the 19th Session brought to an interim conclusion many of the earlier debates within the Council – on the Mid-Term Review, resources required to attain sustainable forest management, continuing progress in Sarawak, trade restrictions, the operation of CITES and the stance of the environmental NGOs.

The reports on the Mid-Term Review and the need for resources provoked lengthy discussion – along predictable lines. There is a constant refrain accompanying any discussion about reporting on sustainable forest management. Reports must not be used to compare or to criticize; reports are mainly to help countries to improve their own management; sustainable forest management is expensive and the resources offered are trivial in relation to the costs; unless progress is demonstrated (presumably through reports), additional resources will not be offered; sustainable forest management brings benefits as well as costs but these are not included in the calculations; some indicators are much more important than others – there should be a core set; all indicators are equally important – none should be downgraded. All these political points continue to be made in every international meeting on the subject! In this debate, they were argued fluently and cogently by many delegates.[11]

In his opening address, the Ghanaian Minister for Lands and Forests, HE Dr Kwabena Adjei, emphasized two points which had for long been the main concern of the producers: that the Mid-Term Review 'should not be construed as carrying the overtones of a threat of judicial sentence carrying a penalty or punishment. Ghana viewed [it] as a chance to reappraise her national experience in sustainable forest management, which meant giving an accurate account of the state of her forests.'[12] And he urged that ITTO should make the connection between progress towards the Year 2000 Objective and the resources required for sustainable forest management.

The Malaysian Minister for Primary Industries, HE Dato' Seri Dr Lim Keng Yaik made much the same point. 'The objective was not to pass judgement or apportion blame. Rather [the Review] should identify gaps and offer constructive recommendations.' Malaysia was taking steps to implement forest management practices based on ITTO's criteria and indicators; from their experience criteria and indicators were best observed and evaluated at the national level.

Many of these questions were negotiable – more money for real needs and good projects; incentives rather than punitive sanctions. But there is one issue of principle over which reconciliation seems impossible, which has to do with the relative importance of indicators. Debates about completeness and inclusiveness rage in every meeting that discusses criteria and indicators, depending upon the speaker's conception of the conditions which are necessary for sustainability.

Mr Mankin (of the Global Forest Policy Project, the only environmental NGO represented at this session) was concerned:

> There was a suggestion … that not all indicators of sustainable forest management needed to be measured. This involved a high risk because it might be that some values considered essential [by] some for evaluation of progress towards sustainable forest management would end up by not being measured. This included the concept of avoiding irreversible impact, particularly with regards to biodiversity.

When pressed to be specific, one high-profile environmentalist has told the author that he would not consider it acceptable for *any* species to be rendered extinct in the course of development. This is, of course, a legitimate point of view for any individual but others would consider it too extreme to be practical. The rubric for sustainable forest management that has been followed by ITTO in relation to biodiversity is that it is legitimate to use certain parts of forest primarily for production; that this is certain to involve some change in biodiversity; that this change is acceptable on two conditions – that adequate provision is made for all a country's species in 'protected areas', and that the change in biodiversity should not be 'undue'. It must be said in defence of this view that natural forest management causes much less change in biodiversity than, for example, conversion to agriculture. But it does cause some.

The Mid-Term Review

The first moment of truth in relation to the Year 2000 Objective came at this session, the date by which all members had been specifically requested to report on progress.[13] The debate was introduced by a consultants' report.[14] The response to the Decision had been disappointing. Only 17 of the 25 producer countries and 14 of the 27 consumer countries had produced a Mid-Term Review. The consultants had used the elements of the IIED report and the ITTO Guidelines as baselines against which to assess the producer reports. In the case of the consumers, most attention was paid to linkages with the management of the tropical forest – reports on trade and international cooperation.

In the producer countries, progress had been uneven and mainly in the field of forest policies, laws and regulations. Advances had been made by a few countries in the involvement of NGOs and local people, in royalty, contract and concessional arrangements, in the development of national guidelines and in the harvesting of a wider range of species. However, in the case of policies, the involvement of local people and NGOs 'does not yet seem to be common'; 'the political will to give greater attention and support nationally to the forestry sector is less clear'; 'few report the formulation of a national strategy to achieve [the Year 2000 Objective]'; and 'only few ... referred to the existence of harvesting rules or to the basis of yield regulation'. Many countries reported the continuing loss of forest cover to unplanned agricultural development. On the positive side, 'the overall impression [was] of an increasing recognition of the basic steps required to establish the policy and institutional base for sustainable forest management, and firm intention, at least within the forestry service, to use the progress made to carry through the intentions to effective implementation'.

All countries reported the existence of a PFE, but there was not enough evidence to judge how much of this was secure or whether the information existed on which to base its sustainable management. There was uncertainty, too, about the representativeness of the TPAs.

Several countries reported greater log production from the forest and many noted the increasing scarcity of prime species. The use of more species and better processing were also reported. Increasing attention was being paid to environmental issues and to consultation with communities.

There was evidence from many reports of the value of the targeted assistance from ITTO and of other assistance being received in connection with TFAP and NFAP activities.[15]

Among the consumer countries, most now had broad policies in relation to tropical forests covering timber, biological diversity, greenhouse gases, etc. Most were either reducing their imports of tropical timber or moving away from logs to processed timber. The volume entering international trade was small and declining. The amount of bilateral aid had increased during the period under review but was 'miniscule compared to the actual needs, as calculated by the Expert Panel'.

It is difficult to summarize effectively the very carefully formulated conclusions of this report. Their flavour is given by the following quotations:

> All countries have correctly addressed the fundamentally important area
> of policy and institutional review and reform. All can show significant
> advances also in other areas, related to the forest resource base, in terms
> of improved information, from remote sensing or intensive inventory,

survey and, in a very few cases, improved knowledge of forest dynamics. However, the progress in actual implementation in the key areas of sustainable forest management, as exercised in each forest management unit, or concession area, is less easy to judge accurately. From the universal expression of the weaknesses in human and financial resources relative to the size of the tasks, it is clear that the current level of field action must be inadequate.

The baseline set in the 1988 ITTO survey ... showed that in many cases the legal basis was adequate, and the most immediate failing was the failure to apply the laws, regulations and controls. It also showed that, while the incidence of well managed timber production forests was extremely low, there were elements of sustainable forest management in all countries. A simple conclusion from this might be that there is no essential change in the overall situation, since there are still extremely few examples of sustainable forest management in practice, and the greatest gap is still between intention and implementation.

One conclusion to be drawn from this study is the impossibility of making generalizations that are universally valid. Effective action, and consequently cost-effective assistance, can only be planned on the basis of analyses *at the country level* of barriers to progress and the preparation of *national strategies* to overcome those barriers.

Resources needed for sustainable forest management

Since the first mention of the Year 2000 Objective, it had been evident that most producer countries would require additional resources to bring their forests under sustainable management as defined in the ITTO guidelines. It was decided in May 1991 to try to estimate the resources required to attain sustainable management of tropical forests by the Year 2000.[16] The Chairman of the expert panel given this task presented his report in May 1992. Total estimates varied widely; resources needed could also vary greatly from country to country; and it was not clear whether the estimates should be for *total* or for *additional* resources needed. The figures from the Expert Panel were for *total* resources. As a result of this discussion, countries were invited to submit their own estimates and another Expert Panel was set up.[17]

By November 1992 reports had been received from only three countries and it was decided to reconvene the panel.[18] In May 1993 guidelines were drawn up to assist countries in making their estimates. Four areas were to be addressed: forest security; an optimal mix of goods and services; efficiency of utilization; and raising public appreciation of sustainable management.[19] After some delay, the matter was taken up again in May 1995[20] in rather a different form: a panel of experts was to 'recommend an approach and methodology for estimating resources needed and cost incurred and to collate estimates submitted by members'. The panel reported in September 1995, basing its discussions on a working paper prepared by consultants,[21] and the report was discussed at the 19th Session in November 1995.

The report of the Expert Panel on the 'Approach and Methodology for Estimating Resources and Cost Incurred to Achieve ITTO's Year 2000 Objective' was presented by its Chairman, Dr Jürgen Blaser. Panel members had first identified the activities and elements needed to cost all the ITTO criteria and indicators; they then selected priority actions from among these and costed them using the figures from the earlier report. They limited their consideration to the PFE and took the view that multiple use was not a necessary element for sustainability; that permanent forest managed for timber production only could be judged sustainable, provided that the timber yield was maintained while other functions and services were not unduly diminished. The definition of 'undue' reduction or damage to other values was therefore crucial.

They concluded that not all the activities identified using the criteria and indicators could be completed within the next five years. Progress depended upon four key areas: security of the forest resource and prevention of unplanned deforestation; implementing sustainable forest management to produce the optimum mix of goods and services; improving utilization to give the greatest possible social net benefit; and improving the social and political environments surrounding sustainable forest management.

Not all the indicators needed to be measured to establish sustainability. Some of them would be decisive, others could be added later. For operational purposes a narrower focus was required; with only five years left until 2000, the panel recommended a realistic approach. The area of forest likely to remain as forest must be secured; harvesting levels must not exceed sustained yield levels; and the impact of harvesting operations must be reduced. With this in mind, seven priority areas had been identified. These were: adopting a forest policy and enforcing legislation; establishing and securing a permanent forest estate; reducing damage done by timber harvesting; training a workforce and accelerating the use of reduced impact logging; limiting harvesting to sustained yield capacity; raising political and consumer awareness of the fact that timber harvesting could be consistent with the sustainability of tropical forests; and concentrating forest research on the analysis and application of existing data and knowledge, and applying research results in management. Their final conclusions are summarized as follows:

> Each producer country should estimate the resources it needs to achieve sustainable forest management. They alone have a full knowledge of all their requirements and are truly capable of identifying all the activities needed and the cost components involved.
>
> ... only four ITTO member countries had submitted assessments... If an estimate of the total resource needs for all producer countries were required, then one basis for extrapolation would be the four country estimates... The Consultants made an extrapolation based on the four country submissions. The Consultants' estimate made along those lines ... amount to US$7,000 million per year. The related benefits and opportunity costs of achieving sustainable forest management have not been estimated.

The priority actions derive from the fact that there are only five years left to achieve the Year 2000 Objective. This time constraint severely limits what and how much can be done... The Consultants estimate that producer countries would need to invest an additional US$2,200 million per year to cover priority actions that need to be completed within the next five years.

The spokesman for the producers linked these two themes – the Mid-Term Review and the need for resources:

the financial resources being made available by the international community were miniscule compared to what were required. The US$2.2 billion per year estimate provided by the working paper was a good indication of what was involved. Agenda 21 had estimated [that] US$30 billion per year was needed, of which US$5.67 billion would have to be provided on concessional terms. So far ITTO had contributed over US$100 million for all projects, 80 per cent of which had come from one country.

Trade

The Malaysian Minister for Primary Industries, Dr Lim, suggested that:

with the broadening of the Agreement the time might be right for all countries to promote timber products, both tropical and non-tropical, as natural and renewable products, particularly in the face of continuing substitution by such 'environmentally unfriendly' products as aluminium, PVC, steel and concrete. Many such products required relatively large amounts of energy to produce, which in turn meant higher emissions of greenhouse gases to the atmosphere. Timber was superior to such products, not only in energy usage but also in terms of renewability, non-toxicity, bio-degradability and insulating properties. To cap this, the beauty of timber was beyond doubt.

He was concerned about the proliferation of third-party certification schemes. He considered that ITTO should examine the role it could play in validating compliance with agreed criteria and indicators.

Mr Mankin showed how difficult it was to get things right. He warned that:

making funds available to overcome consumer bias towards tropical timbers would be widely perceived as little more than a public relations campaign. Indeed many NGOs thought that the Year 2000 Objective was just that. The Organization risked a great deal in investing money in claiming that sustainable forest management had been achieved rather than in actually achieving it.

Mr Geoffrey Pleydell, perhaps in partial answer, spoke on behalf of the UK timber trade:

> It was ironic that the consultants' report had to recommend that ITTO should strengthen its links with the private sector, since ITTO had been born out of a desire to improve international timber opportunities for tropical forested nations. Yet, without the trade there would be fewer jobs, no revenue and precious little reason for looking after forests. The trade was suspicious that the most honourable intentions of CITES were being misused to stop trade in tropical timber species. The timber trade in Britain had fought very hard to defend the interests of the tropical wood producing countries, and part of the defence had been the Year 2000 Objective. Thus, it was becoming increasingly urgent that progress towards this objective was demonstrated. In terms of providing extra funds for sustainable forest management, the ability of the trade was limited and even a 'green premium' would be limited by what consumers were prepared to pay. The UK timber trade had considerable reservations concerning certification systems currently being advocated and alternative solutions should be sought. Evidence of real progress towards the Year 2000 Objective was essential to the trade's fight to protect its interests, which were also essentially the interests of the ITTO member countries.

Environmental NGOs

Mr Mankin[22] was positive about the Mid-Term Review. ITTO had to be able to prove to the world that it was making progress. He suggested that members should concentrate on specific obstacles in each country that needed to be overcome rather than on how long it would take or how expensive it would be. He went on to suggest, supported by Switzerland, that ITTO should cease funding projects that did not support the Year 2000 Objective.

There was some trenchant criticism of NGOs in the course of the meeting. For example, Dr Lim, in his address remarked that: 'ITTO should work with those responsible non-governmental organizations which had the best interests of the forests at heart, but there were other NGOs, competitors and vested interests who served no cause but their own.' He called on ITTO 'to wrest back the initiative from such parties'. Mr Mankin did not rise to the bait; he said:

> he had been asked by several delegates whether he should respond on behalf of the NGOs to criticisms of them made at this Session. However, too much of the Council's time was wasted on tired rhetoric, unfounded accusations and verbal sparring and didn't set the tone that was needed. He would prefer to participate in the Session by offering substance as and when important issues were raised.

13

1995–2000: getting on with the job

'This time, ITTO was determined that its criteria and indicators should be tested and applied'

Council Matters under the International Tropical Timber Agreement, 1994

The new agreement, ITTA, 1994, came into force on 1 January 1997 after some small hesitation on the part of the EU, and the first session under the new Agreement took place in May 1997. Between then and 2000, there was little fundamental change in the activities or priorities of ITTO, except that trade statistics were extended to cover the whole of world trade in timber and not solely that from the tropics. Dr Freezailah gave up his post as Executive Director in March 1999 and Dr Manoel Sobral Filho took his place in November 1999, bringing a new style to the operation of the Organization. The actions and attitudes of the Organization were affected by international developments in the forestry field (outlined in Chapter 11) and were influenced by the approach of the critical year 2000. The forest fires in Southeast Asia and the downturn in the timber trade associated with the Asian economic crisis were also matters of concern.

The main events during this period were: a revision of the Action Plan in 1998; the complete rewriting of the ITTO criteria and indicators, also in 1998; training sessions in their use and attempts to have them more widely applied; the reception of the report of the Mission to Bolivia and the rejection by Council of the proposal that there should be an ITTO mission to the Province of Acre in Brazil; the continuing debate about certification; the much overdue *ex post facto* evaluation of projects; the development of a framework for auditing systems; and activity over transboundary reserves and demonstration forests. Illegal logging became a pressing issue.

Then came the critical year 2000. In preparation for it, a review was commissioned to examine 'Progress towards the Year 2000 Objective'; this took place in late 1999 and early 2000 and was presented to the 28th Session of the Council in Lima (May 2000).[1] Subsequently, a third Action Plan has been prepared and a number of activities have been started, some arising from the recommendations of the Year 2000 Report (see Chapter 15).

The scope of the ITTA, 1994 came up for review in January 2001. Brazil reiterated its view that any future agreement should embrace all types of forests, but it was the general opinion of the Council that the present agreement provided for adequate consideration of non-tropical timber and 'all-forest' issues. So the scope remained unaltered.

Missions

It might be imagined, after the success and influence of the Mission to Sarawak, that there might have been some enthusiasm among members for future missions in spite of the opposition of Brazil. But this was not so; the next invitation was not received until the Cartagena meeting in May 1994. This Mission, to Bolivia, was smaller than that to Sarawak but chosen by much the same procedures. It was led by Dr Ken King of Guyana and paid four visits to the country between October 1995 and August 1996.

The Mission laboured under considerable difficulties, many of which were due to timing. The government was in the throes of implementing fundamental changes in the way Bolivia was to be governed. There was great uncertainty about the future development of many of the country's socio-economic sectors, including forestry. There were many apparently conflicting laws relating to land use and land tenure. A new Forest Law was being prepared; and the many drafts presented to the Mission displayed the very different positions held, for example, by the private sector and environmentalists.[2] There was a great shortage of essential information (eg hardly any growth and yield figures). To cap everything, the Forest Service had recently been dissolved.

The fact that there was no Forest Service made it less than clear who were the Mission's hosts. As a result, although the Mission was able to visit all parts of Bolivia and to meet large numbers of people from the President downwards, there was less opportunity than in Sarawak for the members to decide their own programme; in fact, they did not succeed in visiting any natural rain forests, either managed or unmanaged.

The lack of a Forest Act meant that the Mission was acting in a legislative and administrative vacuum. This did, however, give members exceptional opportunities, by invitation, to discuss with the President and legislators the various drafts of the emerging Forest Law, and there is little doubt that their comments had some influence on its final provisions. This, fortuitously, may have been the most important contribution of the Mission; for the Forest Law of 1996 has had a profound influence on the administration of forestry in Bolivia (see Chapter 14). A very significant step was the establishment of the Forest Superintendency, but there were many other valuable provisions.

The government's request for assistance from ITTO was part of an overall national development strategy, starting in 1993, in an effort to develop an Agenda 21 for Bolivia. Many developments were recent and untried. For example, there were three new Ministries: for Sustainable Development and Environment, for Human Development and for Economic Development, integrated and coordinated by a National Development Council. To support these new institutional arrangements, three new laws had been enacted: the Education Reform Act, the Popular Participation Act and the Capitalization Act.

The Popular Participation Act, in particular, had profound implications for the organization of forestry in Bolivia. It was promulgated in 1994 to permit the active involvement and incorporation of all communities into the economic and political life of the country and to redress the balance between city and countryside. Certain taxes were to be distributed in a proportionate manner throughout the country, and power was to be decentralized to the municipalities, whose administration was, accordingly, much strengthened. This resulted in 308 municipalities of equal standing before the law, and 20,000 grassroots organizations, which had formal access to the municipalities. In addition, the difficulties of the Mission were compounded by the sheer size of the country and the diversity of its forest and ethnic endowment.

The Mission's report[3] gave an account of the forest resources of Bolivia, and reviewed, assessed and made recommendations on all aspects of the forestry sector: forest and land use policies, legislation and strategies; forest administration and institutions; sustainable management of forests for production; forest industries and trade; sustainable catchment management; conservation of biological diversity; the forests and indigenous people; human resources development; and research extension and public information.

The leader of the Mission gave an interim report to the 20th Session of the Council in May 1996 and a full report in November. He mentioned, in particular, the work of the Mission in advising on the new law and the administrative arrangements and regulations that were to accompany it.

His report was well received, even, surprisingly, by the Representative of Brazil:[4]

> Bolivia had invited international cooperation to ensure a better and sustainable utilization of her forest resources. It was to be hoped that the international community would see fit to assist Bolivia in her endeavours.

In fact, two years later, the Brazilian attitude to ITTO missions had changed to such an extent that the Governor-elect of the State of Acre, Mr Jorge Viana, was able to suggest that an ITTO Mission should be sent to the State of Acre. This suggestion was, unfortunately, never followed up by the Council.

In contrast to Sarawak, the report of the Mission resulted in few new projects. This was perhaps because Bolivia had earlier received over US$3 million for the conservation, management and sustained use of the forests in the Chimanes Region of Beni Department. The absence of projects did not, however, mean that no action was taken. The general impression given at a meeting of Andean countries held in January 2000 was that Bolivia was now ahead of all its neighbours in regulating the sustainable management of its forests.[5]

After Bolivia, missions temporarily went out of fashion – until 2000 when Indonesia requested an ITTO Mission. Further, shorter missions have taken place more recently as a result of the recommendations of the Year 2000 Report.

Criteria and Indicators

Since 1992, when ITTO produced its first set of criteria and indicators, many other sets had been produced that covered the field more completely and with greater precision.[6] ITTO had relied on its member countries to adapt the ITTO set to their own use, but only a few had done so. It was clearly time for ITTO to come into line with others and produce a revision that reflected more fully the post-UNCED conception of sustainable forest management. ITTO, therefore, undertook a new exercise to develop a complete set of criteria and indicators that conformed as closely as practicable to this emerging international pattern. The new *Criteria and Indicators for Sustainable Management of Natural Tropical Forests*[7] passed through two expert panels and were agreed in May 1998. The discussions were long and involved and covered well-trodden paths. Should there be some 'core' indicators? What should be the balance between quantitative and qualitative, and between the practical and the ideal? Did they record only changes, or did they imply standards? What were the possible links with certification? What were the most important uses – for the internal auditing of progress, or to demonstrate progress to the outside world (particularly the ultimate consumers)? One point constantly recurred, linked to the question of 'standards' – criteria and indicators should not be used to make comparisons between countries. But how could this be prevented?

There was one difference between the ITTO set and many of the others; it dealt only with natural forests. This was because the expert panels considered that there was a much greater difference between natural forests and plantations in the tropics than there was, for example, in temperate regions, where the distinction was blurred by management and enrichment planting. The expert panels also recommended that a further set should be developed immediately for tropical plantations, but this was vetoed by the US.

This time, ITTO was determined that its criteria and indicators should be tested and applied. A manual was therefore prepared in two parts (one for the national level and the other for the level of the forest management unit)[8] and training sessions were organized in all three continents. This produced a core of experienced operators who could further extend the use of the indicators.

The training sessions led to requests for making the indicators simpler to use, to distinguish between the most important (the 'core' set) and the remainder, and a plea that some be removed because the information was not available or too difficult to obtain. In order to simplify the process, the manuals were replaced in 2001 by a 'Format for Reporting' – essentially a straightforward questionnaire.[9]

The other two problems were more difficult. It had always been the intention that the criteria and indicators would be revised in the light of experience, but it had also been considered vital that they should be *used*; and the plea for revision could so easily be used as an excuse for procrastination. The Year 2000 Report strongly recommended that they should be used immediately for reporting.

As indicated above, the 'core' issue is in principle more difficult to resolve. Some people are convinced that a few indicators are the key to sustainable management (eg the area and security of the forest estate; the balance between harvest and increment); others place their priorities elsewhere (eg biodiversity or public involvement) or believe that all should be included without exception. Indeed, the point is frequently made that it is by the complete set that sustainable forest management is actually defined.

Is there any objective way in which a core set can be distinguished? One possible way would be to separate those indicators which measure the intrinsic capacity of the forest estate to deliver goods and services from the rest – ie the physical and biological variables from the social, the economic and the managerial. The justification for this would be that the last three – social, economic and managerial – all change according to context; they are not the same from time to time. The physical and biological parameters are different in kind; they have a universal and timeless validity. Without an area of forest, productive soil and the capacity for self-regeneration, the potential of the forest to deliver benefits is lost. There remains a very tricky – and perhaps unanswerable – question: how much biological diversity may a forest lose while remaining essentially the same forest in respect of the benefits that it delivers?

Criteria and indicators and reporting remain high on the international agenda. In support of the UNFF, an International Expert Meeting on Monitoring, Assessment and Reporting on the Progress towards Sustainable Forest Management was held in November 2001 in Yokohama.[10] Criteria and indicators were high on the agenda.

The Libreville Action Plan

The first ITTO Action Plan had served to codify the areas which ITTO would wish to encourage and support and it had had some, though rather limited, influence on the annual work programmes. There was need, however, now that the ITTA, 1994 had come into force, to look again at ITTO's priorities for action. The *ITTO Libreville Action Plan 1998–2001* was the result.[11] It was more definite in its approach than the earlier Plan. First, it provided a very anodyne Mission Statement: one cannot help feeling that it could have been more positive.

> The ITTO facilitates discussion, consultation and international cooperation on issues relating to the international trade and utilization of tropical timber and the sustainable management of its resource base.

The Action Plan gives a description of ITTO, its objectives and its structure. It then devotes a section to the Year 2000 Objective, as follows:

> In June 1991, the Council committed itself by Decision 3(X) to what is now known as the *Year 2000 Objective*, which is the goal of having all

tropical timber entering international trade come from sustainably managed sources by 2000. Since then, the Council has approved policy studies and project financing for a number of activities to help member countries move toward this Objective.

The Action Plan listed four key areas for action, identified by the Council in 1993, if progress was to be made towards sustainable forest management. These were:

- Security of forest resources and prevention of unplanned deforestation.
- Production of the optimal mix of goods and services.
- Improvement of the utilization of the resource to give the greatest possible social benefit.
- Improvement of the social and political environment concerning forest management.

The Action Plan next drew attention to the seven priority areas for action by countries that were identified by Decision 8(XX) of Council in 1995 as essential for progress towards the Year 2000 Objective in the short term. These were to:

- adopt a forest policy and apply legislation;
- secure a permanent forest estate;
- apply reduced impact logging;
- train the workforce, including supervisors, in reduced impact logging;
- limit timber harvest to the sustained yield capacity;
- raise public awareness that timber harvesting can be consistent with the sustainability of tropical forests; and
- focus forest research on the analysis and use of existing data and knowledge.

It identified a number of 'cross-cutting strategies' which would apply to the work of all three committees, eg collaboration with other organizations, participation, public relations, training and extension, sharing information, joint ventures, and research. There was one new and significant item at the end of the list: 'to undertake special studies on emerging issues of relevance to the world tropical timber economy'.

Goals were identified for each of the three committees:

- *Economic Information and Market Intelligence*: improve transparency of the international timber market; improve marketing and distribution of tropical timber exports from sustainably managed sources; improve market access and distribution of tropical timber exports from sustainably managed sources.
- *Reforestation and Forest Management*: support activities to secure the tropical timber resource base; improve the tropical timber resource base; enhance technical, financial and human capacities to manage the tropical timber resource base.

- *Forest Industry*: promote increased and further processing of tropical timber from sustainable sources; improve marketing and standardization of tropical timber exports; improve efficiency of processing of tropical timber from sustainable sources.

Within each of the goals, actions were divided into two categories: those which would be carried out by ITTO; and those where ITTO would 'encourage and assist Members, as appropriate'.

The Year 2000 Report

One of the most significant events in 2000 was the Review of Progress towards the Year 2000 Objective, an account of which follows in Chapter 14.

14

Year 2000 Report: the curate's egg

'Guidelines and regulations were all very well; what mattered was the extent to which they were rigorously applied. This was much less certain'

Introduction

There had been so much publicity for the Year 2000 Objective that it was vital for the credibility of ITTO as an agent for change that it should demonstrate to the world that real progress had been made. The year 2000 had come to be looked upon as a watershed for ITTO; upon this its success would be judged. Accordingly, towards the end of 1999, the Council commissioned a study to assess the progress made by countries towards meeting the Objective – and the contribution that ITTO had made to that progress.[1]

The nature of the assessment

Like the 1987–1988 ITTO/IIED survey, this study was designed to analyse the situation, identify the factors that had promoted or impeded progress, and make recommendations. Unlike the previous survey, however, it was *not* based on visits to the countries concerned. Instead it had to rely mainly on country reviews requested by the Council and prepared according to a prescribed format (see Box 10.2).[2]

Members missed an opportunity. None attempted to answer the vital questions at the heart of the Year 2000 Objective – how much of the timber exported came from sustainably managed resources, and what area of forest was under demonstrably sustainable management? As a result, although it was possible to report progress in such matters as legislation, administrative arrangements and consultation, it proved impossible to give an accurate account of any progress towards sustainable management in the forest itself. This was most unfortunate, for it is by action in the forest that progress towards sustainable forest management will be judged.

Interpretation of the Year 2000 Objective

The origin of the Year 2000 Objective has been described in Chapter 6. There were early doubts about the exact way in which it should be interpreted. What was the Year 2000 Objective and what action did it imply?

Since it was first enunciated, the Year 2000 Objective had been formulated or described in several different ways; a summary of its history is given in Box 14.1.

In 1990, the Objective was attached to a target date (2000), to be reviewed in 1995 in the light of progress. At that time, it was known as Target 2000 and was described in these terms: 'The objective is that the total exports of tropical timber products should come from sustainably managed forests by the Year 2000.' But, in 1991, in Decision 3(X) of the ITTC, a much wider remit was added (in italics in the following quotation):

> [a] strategy by which, through international collaboration and national policies and programmes, *ITTO Members will progress towards achieving sustainable management of tropical forests* and trade in tropical timber from sustainably managed resources by the Year 2000.

This is a very significant difference. It seems to denote a change of emphasis from the sustainable management of the forest from which traded timber originates (amounting to some 50 million m^3 from perhaps 100 to 150 million hectares of forest) to the ultimate achievement of the sustainable management of all tropical forests (amounting to some 1300 million hectares) – also, a change of emphasis from 'reaching a target' to 'progress towards that target'.

The first formulation is limited to the sustainable management of those forests from which internationally traded timber originates. For any exported timber or timber product to meet the Objective would require both a confirmation that the particular forest management unit from which the timber came met the conditions of sustainable management and the clear tracing of this timber from the forest to the point of export ('chain-of-custody').[3] This would require detailed examination in the field; in fact, the conditions are exactly those conditions required by timber certification. In this study it was impossible from the evidence provided even to produce an estimate.

The second formulation is much wider. The objective, as well as including that of the first, is no less than the sustainable management of tropical forests – all tropical forests – in ITTO member countries. The implications have been discussed in Chapter 2.

Two conclusions emerge. First, any country can only be said to be managing its forests truly sustainably when the management *of every management unit* is of a sufficient standard to be certified as sustainable. Second, timber can only be guaranteed as coming from a sustainably managed source when the *management of the source* can be certified as sustainable, and the *timber can be traced reliably* from that source.

BOX 14.1 THE EVOLUTION OF THE YEAR 2000 OBJECTIVE (NOW OBJECTIVE 2000)

In the following paragraphs, a brief chronology is given of the evolution of the Year 2000 Objective.

1990

The first reference to the Year 2000 Objective appears in 1990 in the *Draft Action Plan and Work Programme in the Field of Forest Industry* (ITTC(VIII)/D.1) under paragraph 6 (Strategy). This reads:

'Recognizing efforts of producing countries to sustainably manage and utilize their tropical forests and in order to contribute to the achievement of sustainable development in all producing countries, the objective is that the total exports of tropical timber products should come from sustainably managed resources by the year 2000. This target date should be reviewed in 1995 in light of the progress on the implementation of the Action Plan and Work Programme. Therefore, the long term development of appropriate forest based industries in producing countries is the central focus of the strategy.'

1991

In June 1991, the Year 2000 Objective was the subject of Decision 3(X) of the ITTC:

'The ITTC decides to adopt and implement the following strategy by which, through international collaboration and national policies and programmes, ITTO Members will progress towards achieving sustainable management of tropical forests and trade in tropical timber from sustainably managed resources by the year 2000.'

The Decision then went on to describe in detail the steps by which this might be accomplished.

1994

The Year 2000 Objective, in so far as it applied to timber exports, was included as one of the objectives of the ITTO in Article 1 of the ITTA, 1994 in these words:

'(d) To enhance the capacity of members to implement a strategy for achieving exports in tropical timber and timber products from sustainably managed sources by the year 2000.'

1995

A 'Mid-term Review of Progress towards the Achievement of the Year 2000 Objective' was carried out.

1995–1997

Studies were made by panels of experts of the resources needed and the costs that might be incurred in achieving the Year 2000 Objective.

1998

The *ITTO Libreville Action Plan 1998 to 2001* contained a section, as follows:
  'ITTO Year 2000 Objective
  In June 1991, the Council committed itself by Decision 3(X) to what is now known as the *Year 2000 Objective*, which is the goal of having all tropical timber entering international trade come from sustainably managed sources by 2000… '

2000

In Decision 29(XXIX) the Council 'reaffirmed its full commitment to moving as rapidly as possible towards achieving exports of tropical timber and timber products from sustainably managed sources', and renamed it **ITTO Objective 2000**. (The latter formulation escapes from the time limit suggested by the 'Year 2000 Objective'.)

The consultants, therefore, examined and assessed: (a) 'the formulation and implementation of national policies for the *allocation of forest lands* to various uses'; and (b) 'the quality and sustainability of the *management of the production forests* within the permanent forest estate (PFE) paying regard to social and environmental considerations'. They did *not* look at the quality of management of natural forests other than production forests, nor the sustainable management of other areas such as plantations, farm forestry or other non-forest sources of timber. None of the evidence presented had any bearing on these other topics. But, if one takes a strict view of the interpretation of the Year 2000 Objective, both should be included, especially the latter as it is directly relevant to traded tropical timber and to the market in tropical timber.

The report

After an introduction describing the forest cover, forest management practices, and volume and flow of trade in ITTO countries, the report analysed the progress by members. For the producers, this included a country-by-country description followed by a general assessment under the following heads: forest policies and legal framework; institutional framework; forest resource and security of the permanent forest estate; sustained yield management of production forests; utilization; environmental considerations; biological diversity; socio-economic aspects; and international aspects. For the consumers, the treatment was different; attention was concentrated on those actions which might stimulate and assist (or occasionally hinder) progress towards the sustainable management of tropical forests and a timber trade based on sustainably managed sources. The report also briefly touched upon the ways in which consumer countries were approaching the sustainable management of their own forests, as this had a bearing on the equitability of markets.

What follows is an edited and shortened version (without change of fact or interpretation) of the Year 2000 Report.[4]

Analysis of Progress in Producer Countries

Forest policies and legal framework

There had been significant progress in most producer member countries in reforming forest policy and legislation.

In the Asia-Pacific region, all producer countries except Cambodia now had formal national forest policies, most adopted since 1990. Cambodia, for its part, had taken creditable steps to improve forest sector management, including a critical review of forest concession agreements and the regulatory framework. As a result, 12 forest concession licences totalling over 2 million hectares had been revoked. Nevertheless, the policy, legal and institutional environment was still weak, due partly to an acute shortage of trained staff.

Thailand's national forest policy had no legislation to provide for local communities in decision-making. The Thai Forestry Sector Master Plan noted that 'the rural poor who depend on the forest and forest land have no legal rights over the resources'. In Malaysia, the government had reviewed the National Forestry Policy, 1978, amended the National Forestry Act, 1984, and established a National Committee on Sustainable Forest Management in 1994 to map out plans and strategies towards achieving the Year 2000 Objective. This comprehensive body of forest legislation, rules and regulations was now largely in place and would facilitate the implementation of sustainable forest management. Indonesia had taken positive measures along similar lines and had strengthened its regulatory framework, as had Papua New Guinea. A notable development in the Asia-Pacific region was the decision by Indonesia and Malaysia to establish independent organizations to inspect and monitor forest operations for sustainability. In Myanmar, the government had adopted a formal Forest Policy, enacted a new forest law and was involved in an exercise to produce a National Master Plan for the forest sector. The greening of the dry zone in central Myanmar through a community participatory approach had now become a formal programme within a newly created department in the Ministry of Forestry. In the Philippines, the institutionalization of the Community-based Forest Management Programme might complement or even replace the once powerful forest concession companies with local communities.

In Cameroon, a new forest policy had been adopted in 1995. Administrative responsibilities for forestry had been consolidated into a new Ministry of Environment and Forestry, and implementation had been placed in the hands of the National Forest Development Agency (ONADEF). Laws and regulations had been updated and a new classification of forests had been introduced. The Central African Republic had adopted a new policy in 1989 and a revised forest code in 1990 with regulations governing the award and operation of concessions, the conduct of harvesting and the involvement of local people. In Côte d'Ivoire, the seriously deteriorating forestry situation had led to the development of a new forest policy, a master plan for the years 1988–2015 and a substantial reorganization of responsibilities. In Gabon, an extensive set of resolutions affecting forestry had been adopted by the Council of Ministers, a new forest policy had been agreed and a draft law was before the General

Assembly. Changes included a reclassification of forest, revision of the conditions of harvesting licences, and new provisions for community forestry. Ghana had introduced a new Forest and Wildlife Policy in 1995 which had brought striking changes of emphasis. Notable among these were the guiding principles that a share of forest income be reinvested in the resource, that fees and taxes should be treated as incentives for better use, and that the timber industry should move towards a low-volume/high-value trade.

In Bolivia, a new Forest Law had been adopted in 1996. In Brazil, a Forestry Law and Decree had been passed in 1998, and a National Forest Programme had already been partially developed and would be completed during 2000. In Colombia, a new forest policy had been approved in 1996. A Law of Sustainable Forest Development was being prepared in Ecuador which provided for incentives to stimulate the sustainable management of natural forests and plantations. Guyana had implemented a National Forest Policy in 1997 and a national forest plan was being prepared; the President was responsible for forestry, the environment and natural resources and decisions relating to forestry were taken at Cabinet level. A new Forestry Law was being drafted in Peru under the 1993 Constitution and a National Strategy for Forestry Development was being formulated. In 1993, a law had been passed to safeguard the dry forests of the *Costa* and, in 1997, on the conservation of biological diversity and on protected areas. There was a new policy in Panama; a Forest Management Act had been passed in 1992 in Surinam; the policy for Venezuela had been extensively revised under the country's 1999 Constitution; and the 1981 forest policy of Trinidad and Tobago was now being revised.

Most producer countries were signatories to the UNCED-sponsored global conventions and agreements; their obligations and responsibilities to these were reflected in their amended policies and laws.

Several countries had recognized that conservation issues were inextricably linked to the tenure and usufruct rights of rural inhabitants, and had accordingly adopted policy instruments and legislation that strengthened this synergy between production and conservation. For example, Community-based Forest Management formed the basis of law revision in the Philippines, and similar initiatives had been taken in Myanmar, Thailand and India.

The policy and legislative reform undertaken by producer countries broke new ground in several respects. It was a significant shift from timber to resource-based sustainable forest management; it promoted people's participation in resource protection, management and use; and, finally, it focused on the need to protect and conserve natural forests as major reservoirs of biological diversity and as habitats for endangered plant and animal species. A synopsis of the national objectives and forest policies is shown in Table 14.1. National forestry sector reviews had been used as a means to implement these policies, but not all countries had implemented in their reviews the elements they identified in their strategies.

The exact direction of reform depended upon the circumstances of the country in question. In resource-rich countries, such as Congo, Gabon, Guyana, Indonesia, Malaysia, Myanmar and Papua New Guinea, forestry was used to promote socio-economic development with consequent emphasis on forest

Table 14.1 *National objectives/forest policy of selected producer countries in the Asia-Pacific region*

India	Maintenance of environmental stability through preservation and, where necessary, restoration of the ecological balance.	Conserving the natural heritage of the country by preserving the remaining natural forests with the vast variety of flora and fauna – the remarkable diversity and genetic resources of the country.	Checking soil erosion and denudation in the catchment areas of rivers, lakes, reservoirs, for soil and water conservation and mitigating floods and droughts, and for the retardation of siltation of reservoirs.	Checking extension of sand dunes in the desert areas of Rajasthan and along the coastal tracts.	Increasing substantially the forest/tree cover through massive afforestation and social forestry programmes.	Meeting the requirements of fuelwood, minor fodder, minor forest produce and small timber of the rural and tribal populations.	Increasing the productivity of forests to meet essential national needs.	Encouraging efficient utilization of forest produce and maximizing substitution of wood.	Creating a massive people's movement, including the involvement of women, for achieving these objectives and to minimize pressure on existing forests.
Papua New Guinea	Management and protection of the nation's forest resources as a renewable natural asset.	Utilization of the nation's forest resource to achieve economic growth, employment creation, greater Papua New Guinea participation in industry and increased viable onshore processing.						

Thailand	To persuade local people and NGOs to participate in natural resource conservation.	To produce and effectively implement natural resource management plans, particularly for marine and coastal areas, especially mangroves; the previous focus has always been on terrestrial resources.	To introduce budgetary administration for natural resource management.	To eliminate the conflicts of natural resource utilization.	To apply remote sensing and GIS methods to the monitoring and management of resources.	To declare new legislation for natural resource conservation where appropriate.			
Malaysia	To dedicate as PFE sufficient areas of land strategically located throughout the country in accordance with the concept of rational land use to be managed as Protected Forest, Production Forest, Amenity Forest, and	To promote efficient harvesting and utilization within the production of forest for maximum economic benefits from all forms of forest produce and to stimulate the development of appropriate forestry industries	To establish forest plantations of indigenous and exotic species to supplement timber supply from the natural forest.	To promote active local community involvement in various contracts of the forestry development projects and to maintain their involvement in agroforestry programmes. To increase the production of non-wood forest products through	To undertake and support a comprehensive programme of forestry training at all levels in the public and private sectors in order to ensure adequate supply of trained manpower to meet the requirements of forestry and	To undertake and support intensive research programmes in forestry and forest products aimed at enhancing maximum benefits from the forest. To promote education in forestry and undertake publicity and	To provide for the preservation of biological diversity and the conservation of areas with unique species of flora and fauna.	To develop a comprehensive programme in community forestry to cater for the needs of the rural and urban communities.	To foster closer international cooperation in forestry in order to benefit from the transfer of technology and exchange of scientific information.

Table 14.1 *continued*

Research and Education Forest. To manage the PFE in order to maximize social, economic and environmental benefits for the nation and its people in accordance with the principles of sustainable management. To implement a planned programme of forest development through forest regeneration and rehabilitation operations in accordance with appropriate silvicultural practices.

commensurate with the resource flow and to create employment opportunities. To promote a planned development of forest industries towards production of more value-added finished and semi-finished products for local consumption and export. To encourage an aggressive bumiputra participation in the field of wood-based industry in compliance with the government policy.

scientific and sustainable management practices and to supplement local demands and the requirements of related industries.

wood-based industries. To encourage private investment in forest development through the establishment of forest plantation on private lands.

extension services in order to generate better understanding among the community on the multiple values of forests. To set aside specific areas for the purpose of forestry education and other scientific studies.

Myanmar	Protection of soil, water, wildlife, biodiversity and environment.	Sustainability of forest resources to ensure perpetual supply of both tangible and intangible forest benefits for present and future generations.	Efficiency to harness, in a socio-environmentally friendly manner, the full economic potential of the forest resources.	Basic needs of the people for fuel, shelter, food and recreation.	Participation of people in the conservation and utilization of the forest.	Public awareness about the vital role of the forests in the well-being and socio-economic development of the nation.
Indonesia	Policy imperatives To protect ecosystem, soil and water. To sustain multiple goods and services provided by forests to benefit present and future generations. To ensure proper consideration of the views and expertise of all people affected and involved in forest related activities.	Develop the Outer Islands and relieve population pressure in Java and Bali.	Sustainably utilize the mixed tropical hardwood forest and man-made forest for national development.	Develop more productive man-made forest, and convert degraded and unproductive land into productive areas, in order to produce more wood for industrial processing, as well as to restore their environmental integrity.	Generate livelihood opportunities for forest communities and the rural population through the multiple-use management of forest.	Conserve natural resources for the benefit of present and future generations of Indonesians.

industry and trade; but in resource-poor countries a higher priority was given to meeting the subsistence needs of a forest-dependent population. Between these two positions were producer countries, such as Thailand and the Philippines, with modest natural forest resources which they had chosen to conserve rather than sustainably exploit; instead they relied on timber imports. Some countries, too, especially in Africa, displayed great regional contrasts within their boundaries. Despite the differences in their endowment of resource, their social setting and their mix of chosen strategies, most producer countries were giving priority to environmental stability and the principles and practices of sustainable forest management, which they had placed high in their development priorities.

Overall, significant progress had been made in the area of policy and legislative reform in almost all producer countries in all three continents.

The institutional framework

These new policies and legislation had in many cases, especially in Africa and in Latin America, led to consequent changes in the arrangement of ministries and government departments – by establishing ministries responsible for the environment, rationalizing responsibilities and treating the sustainable management of forests in the wider context of national land use. In many countries there was also a strong tendency to decentralize. In general, the more recent arrangements seemed much better suited to deal with present problems.

A serious constraint, however, lay in the shortage of qualified and trained personnel. However good the will, this greatly limited the ability to manage forests effectively. In many cases, it also remained to be seen how effective the new arrangements would be in executing policy.

ITTO studies had emphasized the need for new staff and resources. Malaysia and Indonesia illustrated the scale of the problem. In Malaysia, the Forestry Department and the Forest Research Institute, Malaysia had increased staff strength from 9393 in 1988 to 10,199 in 1999. In contrast, Indonesia reported that they had 9446 technical personnel but needed about 24,700 to ensure sustainable forest management. A systematic assessment was needed from producer countries of the additional manpower required to implement sustainable forest management.

The ITTC had identified reduced impact logging (RIL) as one of the priority actions to achieve ITTO's Year 2000 Objective in the short term. So far, Malaysia had trained 2000 personnel from the private and public sectors, while the Indonesian Concession Holders' Association had organized similar training for about 100 concessionaires. More training (elsewhere) might have been provided but this was not reflected in the country reports. If the benefits of RIL were to be realized, more concerted training programmes would be needed; there was considerable scope and opportunity for sharing information and experiences between producer countries.

Forest resources and the security of the PFE

Very different conventions and terminology were used in different countries. It was, therefore, difficult to make generalizations about the PFE. But, in all countries, there was a growing recognition that it was important to have a productive forest resource that was secure, and that areas should be protected for the conservation of soil, water and biological diversity.

In the Asia-Pacific producer countries the total extent of forested land was about 283 million hectares, with almost 40 per cent in Indonesia followed by India with 23 per cent and Papua New Guinea (13 per cent). In Latin America and the Caribbean it was about 815 million hectares, far the greatest proportion being in Brazil (68 per cent). In Africa, the total extent was 216 million hectares, more than half being in the Democratic Republic of Congo (50 per cent), but with appreciable areas in the Central African Republic (14 per cent), Cameroon and Congo (both with 9 per cent) and Gabon (8 per cent). Details are given in Table 14.2.

Africa

For historical reasons, various systems had developed in Africa, but there seemed now to be convergence towards a common pattern. In Cameroon, for example, since 1995 forests had been classified as Permanent Forests (*le domaine forestier permanent*) and Multiple-use Forests (*le domaine forestier national – non-permanent*). The former comprised ecological reserves, protection and production forests, national parks, recreational and research forests, and community forests; the latter, areas which might at any time be converted to other uses.

In Congo, a national forest estate was planned along lines rather similar to those in Cameroon. The targets were: production forest of 12 million hectares, and protection forests of about 5.5 million hectares. National Parks already covered 2.55 million hectares. In addition, there were over 73,000ha of plantations.

In Côte d'Ivoire, it was a matter of retrenchment. The PFE in theory covered 2.9 million hectares but in fact was virtually restricted to 300,000ha. It was hoped that, by surrounding the remaining highly fragmented natural forests with a buffer zone of plantations, a stable forest estate of 1 million hectares of managed forest, 1 million hectares of unmanaged forest and 250,000ha of plantations might be achieved.

In Gabon, the government strategy envisaged a PFE of 12 million hectares, 8 million hectares of production forest and 4 million hectares of protected areas, all of which would be subject to management plans. Part of the country had already been mapped to put this strategy into execution.

Ghana was the nearest to having a secure forest estate. The forests there were classified for: timber production; permanent protection; convalescence; conversion; and 'not inventoried' – effectively conversion forests. The 'convalescence' forest was degraded and required rehabilitation. Togo had no PFE nor, it appeared, had the Central African Republic.

In most countries, with the possible exception of Ghana, this was still largely a paper exercise. Future security lay in a combination of measures: on one side, the firm definition of boundaries and the implementation of clear management prescriptions; on the other, a demonstration that the security, both of

Table 14.2 *Forest cover of producer countries in 1995*

Region/Developing country	Forest cover (1000ha)	Percentage of land area (%)	Forest area per caput (ha)
Africa			
Cameroon	19,598	42.1	1.5
Central African Republic	29,930	48.0	9.0
Congo	19,537	57.2	7.5
Côte d'Ivoire	5469	17.2	0.4
Democratic Republic of Congo	109,245	48.2	2.5
Gabon	17,859	69.3	13.5
Ghana	9022	39.7	0.5
Liberia	4507	46.8	1.5
Togo	1245	22.9	0.3
Sub-total	216,412	46.5	2.1
Asia & the Pacific			
Cambodia	9830	55.7	1.0
Fiji	835	45.7	1.1
India	65,005	21.9	0
Indonesia	109,791	60.6	0.6
Malaysia	15,471	47.1	0.8
Myanmar	27,151	41.3	0.6
Papua New Guinea	36,939	81.6	8.6
Philippines	6766	22.7	0.1
Thailand	11,630	22.8	0.2
Sub-total	283,418	36.7	0.2
Latin America & the Caribbean			
Bolivia	48,310	44.6	6.5
Brazil	551,139	65.2	3.4
Colombia	52,988	51.0	1.5
Ecuador	11,137	40.2	1.0
Guyana	18,577	94.4	22.2
Honduras	4115	36.8	0.7
Panama	2800	37.6	1.1
Peru	67,562	52.8	2.8
Surinam	14,721	94.4	34.8
Trinidad & Tobago	161	31.4	0.1
Venezuela	43,995	49.9	2.0
Sub-total	815,505	60.1	3.0
TOTAL	**1,315,335**	**51.5**	**0.8**

Source: FAO, 1999.

production and protection forests, was in the interests of local communities. In Africa, it was said, the majority of rural people were more interested in the wealth of biological diversity in the forest than in its potential for timber production. A delicate accommodation was required between these interests.

Asia and the Pacific

The nominal extent of PFE in the Asia-Pacific ITTO producer countries was about 167.7 million hectares (excluding Papua New Guinea and Fiji in Oceania). A system of PFE had yet to be formalized in Cambodia. In India, Malaysia and Myanmar, PFE was constituted through a legal process and procedure. In Indonesia the status of the PFE was becoming more definitive as most of its conversion forest had already been alienated. The Indonesian Government, recognizing that the concession system itself could encourage shifting cultivation, had created a Ministry of Transmigration and Forest Squatter Resettlement in 1993.

In the Philippines and Thailand, all forest lands that were not reserved as TPAs or alienated to other uses were classified as forest reserve. The practice of designating all residual non-alienated land as PFE had several inherent weaknesses. First, people's rights were not accommodated through consultation. Second, there was no sound land use basis for the blanket establishment of reserved forests, because these would probably include lands that are better suited to agriculture or other non-forest use. And, third, it was impractical to protect large tracts of forest lands that were not properly surveyed or managed for a particular purpose, but were merely designated as reserved forests through a mechanical legal device. These areas were not strictly PFE, for they were highly fragmented in distribution and not properly identified or demarcated.

India, Myanmar and Malaysia had a well-established system of PFE due in part to their colonial legacy under the British system of forest administration. The usufruct rights of indigenous people were taken into account through a consultation process required by law. In Indonesia, land set aside for forestry was decided at the local level through a coordinated process of consultation and consensus on land use among interested parties.

Several member countries in the region had also reported increased penalties to thwart timber theft and illegal logging.

Except for the Pacific Island countries of Fiji and Papua New Guinea, most forest lands were State owned. About 97 per cent of the lands in Papua New Guinea were controlled by traditional communities, clans and individuals. Similarly, Fiji had extensive areas of community-owned lands.

The Philippines were taking steps to regularize illegal occupancy in forest lands through a mix of programmes to assist forest occupants in documenting their property rights and in making claims to State forest lands. Programmes included the Integrated Social Forestry Programme, the National Forestation Programme and the Ancestral Land Delineation Task Force.

Finalizing the PFE was still a subtractive exercise in some countries. Higher priority was given to food security and other social imperatives. Nevertheless, many countries had reported a substantial increase in forest lands dedicated to in situ conservation, protection or other environmental purposes. Indonesia, for example, had reported the creation of nine additional national parks totalling 4.5 million hectares over the last ten years. In 1998, the PFE in Malaysia had increased by 13.7 per cent with the inclusion of 1.5 million hectares of State lands, mainly in the State of Sarawak. An estimate of the extent of PFE in the Asia-Pacific ITTO producer countries is shown in Table 14.3.

Table 14.3 *PFE in some ITTO Asia-Pacific producer countries*

Country	Year	Protection forest+	Production forest	Other forests*	PFE areas	Protected
Cambodia[a]	1993		4.73	2.42	7.15	3.40
Fiji[b]						
India[c]	1997	7.39	42.08	12.24	49.47	14.80
Indonesia[d]		33.90	58.50	8.10	92.40	20.60
Malaysia[e]	1998	3.49	10.84		14.33	2.12[j]
Myanmar[f]	1999	0.17	10.40		10.57	1.37
Papua New Guinea[g]						
Philippines[h]	1996	1.38		1.89	3.27	1.34
Thailand[i]	1998				4.88	8.09
Total					182.07	51.72

Sources and explanatory notes
(a) Draft Country Report, 1998; based on 4.73 million hectares of forest concession areas and 2.42 million hectares unallocated forest lands;
(b) Data not available;
(c) MOEF, 1999;
(d), (e), (f), (i) Country reports, 1999;
(g) No natural forests under PFE as per ITTO definition;
(h) DENR, 2000;
(i) In Thailand residual forest lands not under Protected Area System are classified as National Reserve Forests;
(j) 0.33 million hectares within Protection Forest.
+ excludes conservation forests or Protected Areas but may overlap; includes Watershed Forest Reserves in the Philippines.
* includes Unclassed Forests, Conversion Forests, Unallocated Forests in Cambodia.

Latin America and the Caribbean

From the country reports received, the situation of production forests in much of Latin America appeared to be quite serious, although much more promising progress seemed to have been made in relation to protected areas.

In Brazil, most of the potential production forest was on private land. Under new legislation, 80 per cent of any area of private land must be kept under forest and could only be used for timber extraction under a Sustained Forest Management Plan. The tax structure had been revised in 1998, so that forested land was now taxed at the lowest rate.

In Colombia (114 million hectares) there were 46 protected areas (national parks, nature reserves, etc) amounting to 10 million hectares and areas defined as protection forest amounting to 275,000ha. Of the 39 million hectares carrying timber, all but 5 million were limited by accessibility or ecological factors. The potential PFE here had been derived by subtraction and there was no evidence in the national report that it had any statutory protection.

In Ecuador, the draft law stated that 'land under a forestry regime would enjoy protection [by] the police and recommended a system of incentives for sustainable forest management'. It was stated that 2.55 million hectares could be available for immediate harvesting and that the total harvestable area amounted to nearly 3.8 million hectares. As in Colombia, there was no evidence in the national report that any of this potential production forest had any security afforded by statute.

Panama had classified its forests as production, protection and 'special', the last including forest for a variety of uses – scientific, recreational, cultural, etc – but the forests had not yet been delimited either on maps or on the ground.

In Peru, the forest was being reclassified into production and protection forests. It was intended that production should be concentrated in a smaller area, chosen out of a potential productive forest of about 28 million hectares rather than 39 million, thus allowing better administration and control. New arrangements were being made for investment in the proposed PFE. In recognition of the very high biological diversity in Peru, a system of protected areas had been established covering approximately 11.6 million hectares (9 per cent of the country).

Under the 1999 Constitution in Venezuela, the Forest Service would develop action plans to strengthen the classification of areas for permanent forest production through 'plans for classification and regulation of use'. At present, of the 50 million hectares of forest, 16.3 million hectares were assigned for permanent forest protection, but these did not appear to have yet been given any statutory protection.

The situation in the smaller countries in or bordering the Caribbean was somewhat better. Perhaps the most promising of these was Guyana. Here, 13.7 million hectares (out of a forest area of just under 17 million hectares) was classified as State forest, including large areas reserved for research and protection. The various categories of harvesting licences – State forest permissions, woodcutting leases and timber sales agreements – were rigorously controlled.

The forests of Surinam were being classified and regulations for the implementation of management prescriptions were being prepared. In Trinidad and Tobago, 192,000ha (22 per cent of forest lands) were State lands, of which 127,474ha had been gazetted as Proclaimed Forest Reserves.

On the evidence received, only a very small proportion of the production forest in those producing countries in the region that submitted reports enjoyed any measure of security.

Sustained yield management of production forests

The prerequisites for the sustained yield management of production forests are well known and have been discussed in previous chapters. They have been refined further in the recent discussions which have led to the definition of the ITTO criteria and indicators, where some more have been added – such as the use of RIL.

Africa

With the possible exception of 'okoumé' in Gabon, which had been over-exploited in the past, most timber extraction was limited to the harvesting of small volumes of a few species. Lower limits might sometimes be set on the diameters of particular species to be harvested. Although this form of exploitation might do little damage in itself, it often opened the forest to agricultural expansion and poaching. It certainly could not be considered a systematic approach to harvesting based on a knowledge of the dynamics of the forest stand, its AAC and its potential for regeneration. There was a ban on the export of logs from Ghana and, in Togo, all exploitation of natural forest had been suspended in 1988.

There were signs, however, that changes were on the way. A fresh inventory was planned in Cameroon. Concessionaires (concessions covering over 2 million hectares had now been let out) were required to carry out inventories, produce management plans, realize the investments envisaged in their tenders, and apply minimum diameters. The concessionaires had already elaborated their management plans. Much now depended upon the content and quality of the plans and how they were implemented. Management plans would also be required in Gabon. In Congo, management plans were being prepared for three forest units in the north, in which ITTO projects covered forestry statistics and plans for sustainable management and regeneration.

In Côte d'Ivoire, the grant of temporary logging permits had ceased in 1990. It was planned to bring 700,000ha of natural forest under a system of intensive management based on research in Yapo forest. It was hoped that the managed area would be increased to 2 million hectares by 2015. The remaining timber production was from plantations and these were claimed to be under sustainable management.

Ghana had gone further. Resource management plans were required and the management specifications in utilization contracts and in logging manuals were being enforced. Ghana was sufficiently confident of the quality of its management to develop criteria for certification linked to studies of the chain of custody, and to prepare and test standards for sustainably managed forests.

Asia and the Pacific

There had been considerable progress in updating information about the resource. Most member countries in the region could now undertake forest inventories themselves, although there were no national forest inventories in Cambodia and Papua New Guinea. Many countries, too, had made progress in resource monitoring and mapping using remote sensing and geographical information systems (GIS), and had strengthened their capacity to undertake regional and national inventories. In Indonesia, some forest concessionaires were able to map and conduct forest inventories themselves.

In Malaysia and Myanmar, where a system of PFE dedicated to long-term production and protection had long existed and forest management practices had been consistently applied, achieving the Year 2000 Objective was a realistic goal. Indonesia was progressing rapidly to the point where sustainable forest management elements were almost fully in place in some of its production PFE.

In India, the Philippines and Thailand, the deforestation and degradation of productive forests were serious obstacles. The only option appeared to be the forging of beneficial partnerships with local communities in the management and protection of public forests. Compensatory forest plantation programmes were also needed.

With the exception of Papua New Guinea and, to a lesser extent, Cambodia, practically all forest lands under concession agreements in the Asia-Pacific region were covered by management plans. In Myanmar, since 1996, expired divisional working plans had been replaced by district-level management plans. Management plans were constantly being revised in Malaysia. In Indonesia, concession companies were bound by law to prepare management plans covering the period of their agreement. The position was less clear in the Philippines, Thailand and certain Indian states because of logging bans. Several member countries were preparing, and in some cases implementing, forest management guidelines, including forest harvesting codes. There were now guidelines on 'best practice' in various aspects of sustainable forest management. In some cases, improved knowledge of forest dynamics had strengthened the scientific basis for applied silviculture. Generally, therefore, the enabling conditions for sustainable forest management had improved.

Indonesia and Malaysia had criteria and indicators for sustainable forest management – tested in the field, independently evaluated, incorporated in their national forestry standards, and applied in selected forest management units. In addition, there were mechanisms for monitoring and evaluation. Considerable progress had also been made in the certification of forest management at the level of the forest management unit. Indonesia, Malaysia and Myanmar had independent certification bodies. Myanmar had prepared draft criteria and indicators for assessing sustainable forest management, which were being tested in two demonstration sites. These countries had, therefore, made progress towards achieving the Year 2000 Objective; and, given its tradition and commitment to forest management, Myanmar should achieve the Year 2000 Objective in the near future.

Thailand and the Philippines had imposed logging bans in 1989 to check deforestation and to conserve their declining forests. Despite the bans, deforestation and forest degradation continued. These bans could be taken as a 'holding position' to put the whole forestry sector, both demand and supply, into proper focus.

Malaysia and Indonesia had been the first to introduce RIL at an operational scale, but its wider application was hindered by several factors: insufficient trained operatives; the need for considerable planning, commitment and focus; higher harvesting costs; and the need to provide incentives. Some producer countries might also require technical assistance if their traditional logging practices were to be improved.

Latin America and the Caribbean

With the exception of Guyana, Trinidad and Tobago, and Surinam, the situation in those countries that submitted country reports was similar to that in Africa – the extensive low-volume extraction of a limited number of species, restricted, in theory at least, to trees above a minimum girth. The AAC, as in Ecuador for example, was calculated on a national basis; this might, or might not, be related to the potential of the forest types that were being exploited.

But there were plans for improvement. In Bolivia, a large proportion of concessions now had management plans. In Brazil, private forest lands could only be exploited under a sustained forest management plan (PFMA). Colombia reported the formulation of management plans and the application of AAC. The country report from Ecuador was frank in stating: 'The forest stock, in spite of its importance for national development, has been exploited irrationally. The forest sector presents a picture giving evidence of a clear failure in sustainability.' In that country, the reorganization of government structure and the appointment of a *Rector* were an indication of a clear intention to reverse the trends.

In Venezuela, 77 per cent of the volume extracted came from sustainably managed sources and, therefore, met the Year 2000 Objective. Of this, 40 per cent (or about 31 per cent of the whole) came from natural forests under management plans, the rest from plantations and areas with 'annual permissions for selective harvesting'. This was encouraging, but there was no evidence upon what prescriptions the plans and permissions were prepared.

In Guyana, the rigorous conditions upon which licences were issued should ensure sustainable management, and the same was probably the case in Trinidad and Tobago.

General

An estimate of potential yield from the PFE in the Asia-Pacific region, as reported by member countries or based on extrapolated data, is given in Table 14.4 as an illustration of what is potentially possible in all continents.[5]

In so far as the projected sustained yields were potentially much higher than recorded out-turns, it would appear that there is room for increased sustained production from present levels in many countries. But this view may be over-optimistic given the logistical difficulties of organizing harvest operations in remote and hilly terrain, variable quality of the growing stock due to forest degradation, deforestation within the PFE, and other constraints.

Utilization

Africa

The export of timber played an important part in the economies of a few African producer countries, notably Gabon and Cameroon. In Gabon, exports of timber took second place to petroleum, with export earnings of US$250 million from 2.71 million m^3 contributing 9 per cent to the GDP. In Cameroon, exports were valued at US$254 million in 1996/1997 rising to US$422 million in 1998/1999. In all countries, the provision of fuelwood was of great importance; in Côte d'Ivoire, fuelwood consumption amounted to 15 million m^3 per annum.

Much of the timber in the forest-rich countries (Cameroon, Central African Republic, Congo and the Democratic Republic of Congo) came from natural forests, although almost a third of the exports from Congo were of plantation-grown *Eucalyptus* roundwood. Plantations were expected to play a large part in

Table 14.4 *Sustained yield estimates from PFE in Asia-Pacific producer countries*

Country	PFE (million ha)		Potential sustained yield at MAI* 1.5m³ (million m³)		Average annual log production (million m³)	
	Burgess 1988	Present study	Burgess 1988	Present study	Burgess 1988	Present study[a]
Cambodia	na	4.73[b]		7.09	0.00	3.00[b]
Fiji	Data not available					
India	na	34.40	na	51.60	na	15.50
Indonesia	33.87	54.58[c]	50.81	81.87	28.50	31.62
Malaysia	9.09	10.84	13.64	16.26	20.45	33.50
Myanmar	na	10.40	na	15.60	na	2.59
Papua New Guinea	No PFE					3.41
Philippines	4.40	3.27[d]	6.60	4.91	3.43	0.78
Thailand	na		na	0.00	0.00	0.04
Total	47.36	118.22	71.05	177.33	52.38	90.44

Sources: (a) ITTO, 1999b – average production over 1994–1996;
(b) 1999 Country Report – concession area and past average production;
(c) refer to Country Report;
(d) DENR website.
na not estimated in IIED, 1988 study.
* mean annual increment.

the timber supply in those countries where forest resources were scarce or depleted, such as Côte d'Ivoire.

The trade of several countries had been affected by extraneous factors. The war in the Democratic Republic of Congo was having unknown effects; no report had been received from Kinshasa. Both Togo and Congo had been severely affected by civil unrest. In the case of Congo, this had practically put an end to the transport of logs from the natural forest. On the other hand the devaluation of the Currency of Francophone Africa (CFA) had provided a welcome stimulus to trade. In Ghana, the emergence of real estate developers[6] had pushed the domestic price of timber so high that the illegal operation of chainsaws for felling and processing was becoming a serious problem.

Most countries were developing value-added products. Notable among these was Ghana, which had put into effect a series of measures to encourage secondary and tertiary processing; these products were rapidly overtaking lumber in export earnings.

Generally, the installed capacity was more than adequate to deal with production (double), but recovery rates were very low, with 35–38 per cent recorded for lumber and 43 per cent for plywood in Gabon.

Asia and the Pacific

Indonesia was the world's largest producer of tropical plywood and had also the largest rattan cottage industry in the Asia-Pacific region. Peninsular Malaysia had a more diversified forest industry sector than Sabah and Sarawak. With the setting up of a pulp and paper mill in Sabah and additional wood processing capacity in Sarawak, local processing had increased substantially in these two states. Apart from a few modern sawmills managed by the Myanmar Timber Enterprise, the average sawmill in Myanmar was typically small and outdated. Local mills were unable to process efficiently the removals from the forest and, until recently, the AAC for hardwoods had never been achieved.

In 1998, the percentages of domestic felled volumes processed in Papua New Guinea, Cambodia, Malaysia and Myanmar were 19, 73, 82 and 88 per cent respectively. Except for Myanmar, a marked increase in local processing was noted for 1998, probably due to the weak export demand for logs. Over 81 per cent of Papua New Guinea's logs were exported.

Within the Asia-Pacific region, Thailand and the Philippines had become important net importers of tropical logs and, with the lifting of import duties, significant quantities of logs were now being imported by India. In 1998, the volume of imported logs locally processed in Thailand, the Philippines and India were estimated to be 920, 306 and 105 per cent of local production respectively, while Fiji and Indonesia did not export or import logs in 1998.

Most member countries had diversified their wood-based industry, with the emphasis on secondary and tertiary processing, to produce value-added products. The primary wood processing industries in Myanmar, Papua New Guinea and Cambodia could be much improved. The Indonesian Government had stopped issuing new licences for sawmills and plywood mills because of local aggravated log shortages. In Peninsular Malaysia, inefficient sawmills were being phased out by competitive market forces, or by the government if they had ceased to operate for more than two years. Moreover, the shortage of logs had led to increased processing of small diameter logs. The development in rubberwood processing had many parallels to that facing the primary timber industry, but its response to the shortage and quality of its raw material had been innovatively different. As over-mature rubber trees were cut out, the furniture industry had adapted to the conversion of small diameter logs and had also made use of composites like blockboards and finger-jointed components. The export of furniture, using mainly rubberwood components, brought in more than US$1100 million in 1998.

Utilization studies by FAO and ITTO showed 'that almost 50% of the volume contained in trees that are felled is not extracted and utilized'. Two ITTO studies reported that, in East Kalimantan, Indonesia, only 64 per cent of the utilizable felled volume was extracted, while in Sarawak, Malaysia, the recovery rate was only 56 per cent.

Timber supply could be increased by 15–20 per cent if there were better wood recovery in processing and harvesting. Several countries had installed mills for chipboard and medium density fibreboard (MDF), which had resulted in the better use of primary conversion wastes.

Latin America and the Caribbean

The production of industrial timber from all of Latin America in 1997 was about 147 million m^3 per annum, approximately half from non-coniferous forest. Of this, 123 million m^3 were produced by ITTO member countries, of which 66 million m^3 were sawlogs and veneer logs (ie not pulpwood); of this, 46 million m^3 were of non-coniferous timber.

The exports of all ITTO timber products (logs, sawnwood, veneer and plywood) were equivalent to only 6 million m^3 of roundwood. Thus, only about 5 per cent of the production of all industrial timber from the continent found its way into the international market. The exports of pulpwood from the same countries were one-and-a-half times as large. The non-coniferous timber exported had a value of US$1040 million.

The main producer of industrial timber in Latin America was Brazil, with 108 million m^3 in 1997. Half of this was used for the production of sawnwood, plywood or veneer. Brazil, too, accounted for 80 per cent of the ITTO industrial timber for the region, two-thirds of it coming from natural forests; there was no record of the proportion that came from sustainably managed forests. The area of natural forests in Brazil considered suitable for the sustainable production of industrial timber was 245 million hectares, with the assumed potential to supply over 200 million m^3.

In spite of the fact that most of its production was for the domestic market, the main exporter in the region was Brazil (the equivalent of 4.7 million m^3 in 1997); the accumulated exports of all the other countries in the region were only equivalent to 1 million m^3. Bolivia and Guyana exported the highest proportion of their production, about 20 and 50 per cent respectively. No timber was exported from Peru.

The export of logs from natural forests was banned by all countries except Guyana and Surinam, but this tacit agreement had recently been broken by Ecuador under the influence of a structural adjustment programme.

The recovery rate in Ecuador was reported to be 44 and 51 per cent for chainsaw and sawmill respectively; and in Panama 50 per cent for sawnwood and 85 per cent for plywood.

Environmental considerations

The most serious environmental damage associated with timber harvesting was soil disturbance and canalization of water caused by bad road design and construction, and by skidding. The remedies for this were twofold: the introduction, implementation and enforcement of strict engineering specifications for road design; and the use of RIL. There was no specific mention in any of the country reports of engineering standards or of measures to protect the forest against chemical pollution.

Many countries now had national environment-related legislation, where the potential adverse impact of development must be analysed through EIA, but this did not always apply to major forestry operations. Examples were: Cameroon, India, Indonesia, Malaysia, the Philippines, Panama, Peru, and Trinidad and Tobago. Legislation was in draft for Gabon and had been recommended for Togo. EIA guidelines were being prepared in Ghana for the wood industries, and in Guyana the issue of a licence was conditional on the preparation of an EIA statement.

The 1997/1998 Indonesian forest fires had caused heavy air pollution in neighbouring countries. The scale of this calamity had brought into focus the need for collaboration among affected countries in improving forest fire prevention and management capability in the region. A national guideline on integrated forest fire management had been prepared by an ITTO-assisted project.

Biological diversity

A number of issues arise in relation to sustainable forest management. Some of these concern protected areas, others the management of production forests. In relation to protected areas, the most important are:

* the distribution, size and representativeness of the protected areas;
* the extent to which they incorporate special areas (eg those that are very rich in species or the habitat of endangered animals or plants);
* the nature of the land use between the protected areas;
* the ways in which the purposes of protection may be harmonized with local people; and
* the monitoring of their effectiveness.

In almost all countries, great advances had been made in extending the coverage of protected areas and, in a number, there had been innovative approaches to harmonizing protection with the interests of local people. In all regions, wild forest species had been collected, harvested or hunted. They were, perhaps, particularly valued and valuable in the African tropics.

Many of the protected areas, however, were not adequately managed; and management was crucial where pressures of population and of other uses were growing around them. There was also little evidence that the harvesting of wild species was being conducted in a sustainable manner, especially when they had become commercialized.

Rising population pressure, grazing and fire were some of the problems facing management and protection in India and in many other member countries. In Indonesia, a buffer zone had been created between the national parks and the populated areas, and, in Côte d'Ivoire, even between potential production forests and farmland.

It is valuable also to protect samples (in as undisturbed a state as possible) of all the forest types that are being exploited for timber, for two purposes: as genetic reserves, and as baselines against which changes in the managed forest may be measured and assessed. The extraction of timber is certain to induce changes which are likely to be more extreme the greater the intensity of forest harvesting. In this context, in Malaysia, Virgin Jungle Reserves (VJRs) had been established in different forest types and ecotypes within the system of PFE in the country.

Although much of the conservation of biological diversity can be accomplished in well-managed protected areas, some consideration must be given to measures within production forests. There is a set of ITTO guidelines on this topic. There was no evidence anywhere in the country reports that this question was being addressed. The correct place to do so would be in the specifications for management plans.

Socio-economic aspects

An increasing trend was the greater involvement of NGOs, community organizations and local people in managing public forests. This was reinforced by the widely held view that, without peoples' commitment and involvement, the protection of public forest lands was almost impossible given the limited manpower and financial resources of forestry administrations in most countries. Encroachment into forests was a major social problem in many countries in all regions. In India, about 1.5 million hectares of forest land had been illegally occupied for agriculture and other uses.

The evolution of different community-based forest management models, with common emphasis on sharing of forest resource benefits and services, had given rise to different types of tenure instruments in member countries. In the main, with the exception of the Pacific Island countries, they drew their legitimacy from the government, for it was government rather than the community that instituted them. In reality, the paramount giver and enforcer of rights to forest resources was still the state forestry administration.

Sources of wood supply outside the PFE were very important in most of the drier countries. Even in Indonesia, about 20 per cent of local wood demand was met from outside the forest. Indonesia, Malaysia and Thailand had considerable areas of rubber plantation; there, rubberwood – a by-product of the estate replanting programmes – was a significant component of the tropical timber trade. The demand for fuelwood could be greater than that for industrial wood. For example, in India, fuelwood constituted 70 per cent of its total production.

The conversion of small dimension logs and the domestic use of a wider range of species reflected more realistically the species-richness and heterogeneity of tropical forests. As more primary forests were logged, future supplies would come from second-cut or residual forests where the dimension of the harvested logs would probably be much smaller. There was great scope, as in Bolivia, for the enrichment of degraded areas or secondary forest with indigenous fast-growing species. In Bolivia, there was also now a market for the fruits and seeds of such species and a manual was being prepared on ways in which they might be identified, reared and established.

Non-wood forest products (NWFPs), like bamboo, resins, rattan and many others, are economically important not only locally, but also sometimes form the basis of major cottage export-oriented industries and a considerable export trade. Indonesia and Myanmar reported that NWFPs, comprising rattan, bamboo, resins, guano and edible bird's nests, had contributed significantly to foreign exchange earnings, rural employment and trade.

In Indonesia, the forest and forest industry sector was used to promote economic growth, improved social welfare and employment in rural communities. Accordingly, concessionaires were required to provide social infrastructure like roads, forest villages and other supporting social amenities to rural communities. In Papua New Guinea, forest concessionaires provided a package of rural infrastructure like schools, clinics, houses, roads and bridges, etc, and disputes often arose, where the standard of facilities provided was not up to the expectations of rural forest owners.

The Position in Consumer Countries

Reports were also requested from consumer countries and were received from the EU (and separately from the United Kingdom and Sweden), Norway, Switzerland, the US, Australia, New Zealand, Japan, China, Korea and Nepal.

All consumer countries had stated their commitment to the sustainable management of their forest resources and had developed forestry sector plans; they were also parties to the UN conventions. From the reports received, it was evident that all the developed consumers were using criteria and indicators for sustainable forest management.

A few also reported the development of timber certification schemes. There was a danger that the proliferation of different timber certification schemes in consumer countries might generate uncertainty and curb market access for producer countries instead of, as intended, promoting sustainable forest management practices.

Interactions

This section[7] was intended to examine the interactions between producer and consumer countries, and between both of these and ITTO, specifically in relation to the Year 2000 Objective. Producer countries provided little information in their answers about the two most important questions: 'how adequately has international assistance targeted the Year 2000 Objective in focus and in amount?'; and 'what influence have trade policies, such as incentives, tariffs and bans, had on progress towards achieving the Year 2000 Objective?'. Almost all concentrated on the amount and kind of assistance received, although they also mentioned some of the cross-currents caused by fluctuations in trade policies and in the international market.

The initial concern that the pressure for timber certification was a front for the impositions of bans on the import of tropical timber had now abated. It was becoming recognized that timber certification was going to come and that it would be as well for producer countries to be in on the act.

The consumer pressure being placed on the trade from retailers and local government authorities had been sufficient for a number of tropical producing countries to consider it important to develop their own timber certification systems. It had also begun to be recognized that the establishment of the standards of performance necessary for certification would have a beneficial effect on their own forest management. Thus, an interest in timber certification was growing in certain countries on all continents and in some timber organizations: Cameroon, Ghana and the ATO in Africa; Ecuador and Surinam in the Americas; Malaysia, Indonesia, Myanmar and the Sarawak Timber Association (STA) in the Asia-Pacific region.

One notable feature in the Asia-Pacific region was the high degree of intra-regional trade among ITTO member countries. Japan, China and Korea and, to a lesser extent, Thailand and the Philippines were the major importing countries. Hence, any changes in policy and economic conditions in these countries would have a large influence on the pattern of log and timber products trade in the region.

Generally, the trend towards greater liberalization in world trade would have a positive effect on trade in forest products. While the importing countries had largely removed trade barriers and tariffs on primary and semi-finished wood products (eg sawlogs and sawn timber), there were still many non-tariff barriers imposed on value-added products. For example, the EC maintained a plywood import quota. In the Japanese market, tariff rates ranging from 4 per cent to 8 per cent were applied to Malaysian sawn timber, veneer, plywood and mouldings. Despite the yearly reduction in tariffs, export of plywood and other forest products to the Australian market still incurred relatively high tariffs. An import duty of 5 per cent had been imposed under the general preferential tariff. These trade restrictions translated into higher processing costs for producer countries relative to those in developed economies.

The industry had also been affected by a growing number of regulations, policies and conditionalities placed on tropical timber imports. To the extent that these new regulations were based on environmental and sustainable forest management requirements, they were basically different from tariff and quota barriers that were meant to protect the industry, social security and balance of payments of importing countries.

Labour and environmental clauses had been added to the EU's generalized scheme of tariff preferences (GSP), which came into effect in mid-1998. Countries complying with International Labour Organization (ILO) and/or ITTO standards for sustainable forest management might qualify for additional tariff reductions in addition to normal GSP preferences.

In the Asia-Pacific region, Japan would continue to be the region's largest importer of forest products. Therefore, changes in Japan's consumption preference and pattern, economic growth or the extent of harvesting of Japan's forest resources would significantly influence trading patterns in the region. Given their large population and growing demand, China, and to an increasing extent India, would also have an effect on regional trade balance and direction. Moreover, some former producer exporting countries in the region were now net importers, eg Thailand and the Philippines. At the same time, rising domestic demand in the producer countries might also reduce the amount of timber available for export.

Several countries had also imposed a ban on various forms of logging in a bid to save their remaining natural forest assets. The opportunities and costs associated with such a policy had generally not been properly taken into account in the forest sector strategy of these countries. Wood imports, as an alternative source, should in no way compromise national commitment and investment into forest resource creation and renewal programmes, nor should it be a disincentive to rural wood production.

The trend in the producer exporting countries had been to produce more value-added exports through improved and expanded secondary and tertiary wood processing. The furniture trade in Ghana, Malaysia, Indonesia and the Philippines had contributed significantly towards job creation and foreign exchange earnings. Some wood industries in Japan were being relocated to primary producing countries in the face of decreasing log imports.

A balance should be struck between maintaining the competitiveness of domestic industries on the one hand, and managing the consequences of trade liberalization in a globalized environment on the other. For example, the elimination of the Indonesian Panel Products Manufacturers Association (APKINDO)'s monopoly over plywood exports and reduction of export taxes on logs and rattans based on International Monetary Fund (IMF) loan conditions would affect the competitiveness of Indonesia's wood industry. A summary of export taxes and bans on tropical timber of the ITTO producer countries in the Asia-Pacific region is summarized in Table 14.5.

For their part, all the consumer countries reported a reduction or a phased reduction of import tariffs on timber and timber products, especially those from developing countries, while the EU had provided double tariff preferences to countries which complied with specific ILO conventions and/or whose tropical wood products met the ITTO criteria for sustainable forest management.

A summary of trade barriers perceived to operate in major ITTO consumer countries is given in Table 14.6. The table shows that a higher level of tariff and duties for processed products was still being applied in several consumer countries. Any such tariff escalation between raw log and processed timber products could have considerable implications for developing country exporters, because higher relative tariffs for processed products, would discourage the development of value-added wood processing in exporting countries.

The developed consumer countries had provided substantial[8] technical and financial support, both bilaterally and multilaterally, to enhance the capacity of the producers to move towards the Year 2000 Objective. They had also pledged to increase their financial contributions, and one country reported that its overseas development assistance had already exceeded the target of 0.7 per cent of Gross National Product (GNP).

Conclusions

The conclusions were summarized as follows. *They describe the position as found in 2000.*

Producer countries

From the material available, it was only possible to make a preliminary and approximate assessment of the progress made by producer countries towards achieving the Year 2000 Objective; a more thorough assessment would have

Table 14.5 *Export taxes and bans on tropical timber in producer countries in the Asia-Pacific region, 1999*

Country	Tax rate/export policy	Constraints/incentives/disincentives
Cambodia	(a) Ban on log and sawnwood exports; (b) 12 forest concession agreements revoked (about 2.7 million hectares) due to non-compliance.	(a) Reduced royalties/taxes as incentives to promote local export-oriented processing; (b) Shortage of trained manpower; (c) Political stability vital to overall development.
Fiji	(a) Proposed ban on green sawnwood export to promote local processing; (b) Increased production of lesser-known species planned; (c) About 4.7 million hectares under concession agreements.	(a) Lack of trained technical expertise in wood industry; (b) Reduced duty on sawmill machine import; (c) Concern over increased cost of timber certification.
India	(a) Import duties on logs, firewood and charcoal lifted in 1997; (b) Several states imposed a ban on green logging in the 1980s.	(a) Availability of cheap imports a disincentive for investment in trees, especially private farm forestry growing *Eucalyptus* chips for local pulp and paper industry.
Indonesia	(a) Export tariff rates for logs, lumber and rattan will be reduced finally to 10% in Dec 2000; (b) Introduce resource rent tax and IMF-driven reforms; (c) Import tariffs on timber products further reduced in 1997.	(a) Dismantling of APKINDO, the Indonesian plywood cartel, may improve competitiveness in plywood manufacturing sub-sector.
Malaysia	(a) No duty on log, sawnwood and face veneer imports; (b) 25% import duty on core veneer and mouldings; (c) 45% import duty on plywood; (d) Log export ban in Peninsular Malaysia; (e) No plan to expand Sabah wood processing industry.	(a) Sawn timber and log imports from Indonesia may increase; (b) Higher royalties/taxes for exported logs to promote local wood processing industry.
Myanmar	(a) Only Myanmar Timber Enterprise can export teak logs and undertake teak logging.	(a) Modernization of private/public sector wood processing industry required to improve productivity and efficiency; (b) Monopolistic role of Myanmar Timber Enterprise vis à vis creation of market-driven private sector investment into wood processing industry and trade needs to be reviewed.

Table 14.5 *continued*

Papua New Guinea	(a) Import tariffs of 30% for logs and sawnwood, and 55% on plywood and veneer in 1999; (b) Log export will be phased out.	(a) Local industry will face increased competition due to the reduction of plywood/veneer tariffs from 100%; (b) Log export badly affected by Asian crisis; (c) Local forest industry may not be efficient enough to compete in export market.
Philippines	(a) No tariff on log import, 3%/15% on coniferous/non-coniferous sawnwood and veneer, and 25% on plywood; (b) Logging ban on virgin forests introduced in 1989; (c) Only plantation grown logs and sawnwood can be exported.	(a) Several incentives to promote investment into forest plantations; (b) AAC severely reduced as future production will come from residual, second growth and not from virgin forests; (c) Eco-labelling of forest products may affect trade.
Thailand	(a) 1996 reduced tariffs for ASEAN forest products: logs – 5%, sawnwood – 10%, veneer – 20%, plywood – 20%; (b) Logging ban introduced in 1989.	(a) Log imports from Malaysia, Myanmar, Cambodia and Laos will continue on increasing scale; (b) Local wood processing industry restructuring due to shortage of raw materials – from primary to secondary/tertiary processing.

Source: Mainly from ITTO, 1999b, and Country Reports.

required more intensive survey. Nevertheless, it was possible to record very considerable improvement over the situation recorded in 1988 or in the Mid-Term Review. The most striking advances were in the fields of policy and legislative reform.

Many of these reforms were initiated by ITTO; but they had been spurred on by the demand in the international market place for timber and timber products to come from sustainably managed sources, by privatization and trade liberalization, by greater awareness of environmental and conservation issues and by the need to forge enduring partnerships with local people in resource management. Also, many countries had been considerably influenced by changes in the international field since UNCED.

As a result, almost all countries had developed new policies for their forests and forestry, often within the framework of wider land use or environmental policies; and they had supported these policies by enacting new forest legislation. In doing so, they had provided the conditions in which further advances towards sustainable forest management had become possible.

Many countries now had national environmental legislation. The potential adverse impact of development must be analysed through EIA, but this did not always apply to major forestry operations.

Table 14.6 *Tropical timber trade barriers in consumer countries, 1993*

Country	Product	Description
Australia	Sawnwood	2–7%, depending on species and origin, phased annual reduction since 1995.
Canada	Logs, sawnwood, veneer	None.
	Plywood	8–9% import tariff, depending on species.
China	Logs, sawnwood, woodchips	No import duty since January 1999.
	Veneer	Reduced from 12 to 5% from January 1999.
EU		Import quota on plywood.
France	Logs, sawnwood	None.
	Plywood	10% (except waivers under GSP, Lome).
Netherlands	Logs	None.
	Sawnwood	0–2.5% import tariff, depending on species.
	Veneer	4–6% import tariff, depending on species.
	Plywood	10% import tariff, depending on species.
Portugal	All	None.
UK	Logs, sawnwood	None.
	Plywood	9.4–10%, depending on species.
Japan	Veneer	Tariff base rate 15% (subject to GSP scheme), to be reduced to 5% by 1999.
	Plywood	Tariff base rate 17–20% (subject to GSP scheme), to be reduced to 8.5–10% by 1999.
New Zealand	Sawnwood	None.
	Finger-jointed timber	7%.
	Tropical plywood	6.5–7%.
	Tropical veneer	6.5–7%.
Norway	All	None.
Korea	Logs	2% import tariff.
	Sawnwood	5% import tariff.
	Veneer	5% import tariff.
	Plywood	8% import tariff.
US	All	None (GSP scheme).

Source: ITTO, 1996, and China and Canada Country Reports, 1999.

It was possible to affirm that significant progress had been made in policy and legislative reform in almost all producer countries in all three continents. Parallel developments had taken place in the consumer countries.

This legislative reform had frequently been followed by a reorganization of administrative arrangements, and a restructuring of ministries and government departments, often by establishing ministries responsible for the environment, by rationalizing responsibilities and by treating the sustainable management of forests in the wider context of national land use. There had been a move, also, in a number of countries, to devolve substantial responsibility for implementation to regional or local authorities.

Many countries, too, had developed new strategies or master plans for forestry, frequently based on the results of remote sensing, GIS technology and new forest inventories.

However, there was not yet strong evidence that the strategies were being acted upon. The strongest reasons for this, advanced in almost all country reports, was the shortage of qualified and trained personnel, and of finance. The impression given in these reports was that the will to implement was there; the means were lacking.

Considerable progress, too, had been made in most countries in establishing a PFE. This was usually on State land, but, where private owners were involved, the same end was being achieved by providing more security of tenure, by financial incentives and, sometimes, by legislation. The much greater degree of consultation with local communities was having some effect in gaining local support for sustainable forest management and reducing encroachment and damage. Nevertheless, illegal logging and poaching still remained a problem in many countries.

In this connection, many countries reported a substantial and welcome increase in forest lands dedicated to conservation, soil and water protection and other environmental purposes as part of their PFE. There had also been innovative approaches to harmonizing protection with the interests of local people. In all regions, wild forest species were traditionally collected, harvested or hunted; these were perhaps particularly valued and valuable in the African tropics. The case for the conservation of biodiversity, if properly argued, was one that appealed to many local people.

Many of the protected areas, however, were not adequately managed; and management was crucial where pressures of population and of other uses were growing around them. There was also little evidence that wild species were being harvested in a sustainable manner, especially when they had become commercialized.

Finalizing the PFE was still a subtractive exercise in some countries. In the course of land use planning, land had often been allocated to forestry only when it was not required for any other apparently more urgent purpose. However, more attention was now being given to comprehensive land use planning which took all aspects of the national interest into account. It was becoming recognized that sustainable forest management was not a 'production' accounting exercise that merely balanced potential yields, secured under different management regimes, against aggregated out-turns, without regard to the nature, form and extent of the resource and to the objectives identified for different parts of it.

However good the policies, laws and administrative arrangements, the success of sustainable forest management has to be judged by results in the forest. The authorities in most countries were now fully aware of the quality of management that they should aim to implement. Many were using the ITTO series of guidelines and the ITTO criteria and indicators (also those of other processes) to develop national guidelines and national criteria and indicators. Some were going further to develop standards for forest management and investigating the possibilities of timber certification. Most countries were following the logical sequence from idea to implementation. Some were much further advanced than others, but all seemed to be moving in the right direction.

Guidelines and regulations were all very well; what mattered was the extent to which they were rigorously applied. This was much less certain. The reports had almost nothing to say about the extent to which timber harvesting, road building and stand management in individual forest management units was following the prescriptions laid down for them.

Reports of achievement at the national level could not, alone, give the full picture. They needed to be supplemented by an account of the extent of effective implementation at the forest management unit level, for this was the true measure of progress towards achieving the Year 2000 Objective. It was only by verification in the field that such information could be provided. The consultants believed that it was vital that there should be field verification, at least of a random sample of forest management units, if countries were to provide convincing evidence of progress. This kind of information was also required by the countries themselves in order to identify the areas in which further improvement was needed.

Although the national importance of sustainable forest management was now widespread in government and in a small sector of the population, it was necessary to disseminate the message much more widely, especially among the concessionaires, the timber industry, forest workers and the farming and other communities living in and near the forest. They must become convinced that sustainable management of the forest was in their own best interests and, wherever practicable, be involved in the process.

A sustainable timber harvest depended upon keeping the level of harvesting at or below the AAC and allowing sufficient advance growth and enough seedlings and saplings to ensure a future crop. The extraction of timber in many countries was still based on rule-of-thumb minimum diameter limits or guesswork. Many countries did not have enough information (from growth and yield studies, studies of the biology of the principal timber species, etc) to provide a sound basis for sustainable harvesting.

The most serious environmental damage associated with timber harvesting was soil disturbance and the canalization of water caused by bad road design and construction, and by skidding. The remedies for this were twofold: (a) the introduction, implementation and enforcement of strict engineering specifications for road design; and (b) the widespread use of RIL. There was some evidence that the problems were being taken more seriously in some countries; for example, RIL was becoming more widespread in the Asia-Pacific region and was being encouraged in Latin America.

If large areas of forest were to survive through sustainable harvesting and use, this could only come about through making this management profitable to all concerned. Any measures to increase the value of forest goods and services would increase the chance that management would become sustainable. Also, the chances of success would become greater, the larger the number of people who had a stake in sustainable management. Several measures were being taken to increase value and to widen the range of stakeholders.

It was important that concessionaires and timber companies should have a greater role in ensuring their own future by being given wider responsibility for preparing their forest management plans and in developing guidelines and standards for their operations. There were signs that this was taking place in some countries.

As more primary forests were logged, future supplies would come from second-cut or residual forests where the dimension of the harvested logs would probably be much smaller. More attention was now being given to the conversion of small dimension logs and the domestic use of a wider range of species, and to the enrichment of degraded areas or secondary forest with indigenous fast-growing species.

Countries were now taking more seriously the harvesting of the wide variety of NWFPs for local use, as the basis of cottage industries and for export. They were also making use of the potential of their forests for ecotourism.

There was an encouraging trend in many countries to produce more value-added exports through improved and expanded secondary and tertiary wood processing. There were strong market arguments, too, for developing a demand for high-value timber species and favouring them through silvicultural treatment.

Certification systems were being developed and tested in a number of producer countries. Several of the representations that the consultants received from timber trade organizations urged ITTO to take more of the initiative in certification.

The present trend towards greater liberalization in world trade had had a positive effect on trade in forest products. While the importing countries had largely removed trade barriers and tariffs on primary and semi-finished wood products (eg sawlogs and sawn timber), there were still many non-tariff barriers imposed on value-added products. The industry was also affected by a growing number of regulations, policies and conditionalities placed on tropical timber imports. To the extent that these new regulations were based on environmental and sustainable forest management requirements, they were basically different from tariff and quota barriers that were meant to protect the industry, social security and balance of payments of importing countries.

Six countries appeared, in 2000, to be managing some of their forests sustainably at the forest management unit level to achieve the Year 2000 Objective, while others were moving in the same direction. In all of them, however, there were still problems of full implementation in the forest. The six were:

1   *Cameroon*: The PFE had been defined on the basis of inventory. Arrangements were being made to respect the interests of local people. A management plan was compulsory. Knowledge of growth and yield of the principal species and of their regeneration potential needed to be strengthened to ensure sustainable forest management.
2   *Ghana*: Many elements of sustainable management were in place. The PFE was now established and secure. There would be specifications for sustainable management in management plans and these would be enforced. There was a package of incentives for sustainable forest management. There were arrangements for a proportion of profits to be reinvested in forest management. A process for certification and standards for certification was

being developed. The profitability of the industry had been increased through the development of tertiary processing. The prospects were good, but they had yet to be realized.

3   *Guyana*: There was a secure PFE. Forest concessions were awarded under stringent conditions and by a penetrating process which required a management plan, an environmental impact assessment and an environmental management plan. The standard of implementation still needed to be confirmed.

4   *Indonesia*: All the key sustainable forest management elements were now in place. The country's PFE was secured, based on land use priorities identified through local broad-based consultation processes. National forestry standards, a certification body and supporting activities had been institutionalized. Still to be addressed was the wider application of sustainable forest management practices to the PFE, supported by improved capability and expanded capacity, and improved enforcement.[9]

5   *Malaysia*: The country had a system of demarcated PFE, complemented by well-managed networks of TPAs and VJRs. Forest management was under control and silvicultural treatments were adequately funded, backed by improved knowledge on forest dynamics and research oriented towards sustainable forest management. Management plan prescriptions were implemented and AAC limits followed. Greater involvement of interested parties needed to be further promoted. The country had a clear strategy towards achieving the Year 2000 Objective.

6   *Myanmar*: The country had a system of PFE that was managed under approved working plans. Criteria and indicators for sustainable forest management had been drafted and two demonstration sites established. Key steps had also been taken to develop a certification procedure. More attention to the sustainable forest management of the non-teak hardwood resource was required, and specific technical assistance was needed.

It was difficult to generalize about producer countries as a whole; they had such very different endowments of forest, national wealth and densities of population. While the steps to achieve sustainable forest management might be similar for all countries, the particular factors that limited progress were likely to be country specific. In fact, the 'priority activities' for one country might not be the 'priority activities' for another. This had implications for the best ways in which the situation might be influenced by ITTO, aid organizations and potential investors. Unfortunately, very few countries had identified specific constraints in their reports, apart from the lack of adequate finance and trained manpower.

Producer countries that had made the most progress in achieving the Year 2000 Objective were those with a good resource endowment, who were major exporters of timber and timber products, and had the possibility of establishing a PFE. In other countries, it was a major challenge to protect the PFE in the face of mounting population pressure, large unmet demands for forest products and services, and the growing importance of timber production outside the forests. For these countries, establishing demonstration forests to test the practicability of sustainable forest management according to the ITTO processes might be a good start towards achieving the Year 2000 Objective. Alternatively, their best approach might be to derive their timber from sources outside the PFE, such as farm forestry or private plantations (or from imports).

A review of country submissions indicated that demonstration forests had contributed towards the development and adaptation of approaches to sustainable forest management in several producer countries, such as Malaysia and Myanmar. Continued technical support in the setting up of such demonstration forests within the major productive forest types would facilitate and enhance the progress of member countries towards achieving the Year 2000 Objective.

Finally, it was vitally important that the world at large should know about the positive changes that were taking place. Although slow, movement was consistently in the right direction. Much more needed to be done about publicizing these advances in the interests of the countries and their people, the forests and the trade. Very few of the country reports had responded to the clear invitation from ITTO to explain the significance of the changes they were making or to identify clearly the precise nature of the difficulties they were encountering. None of the reports would have set the world on fire! This was a pity and a lost opportunity. Many of them did, however, make clear that the assistance they were receiving – from ITTO, in bilateral aid from consumer countries and from multilateral aid – was reasonably well targeted in relation to the Year 2000 Objective. It would be beneficial if more good project proposals were to be presented addressing the most important constraints in any country to achieving the Year 2000 Objective and that these should be funded.

Consumer countries

It was clear from the country reports that all consumer countries of ITTO were committed to the sustainable management of their forest resources. Those from the developed world had all incorporated criteria and indicators for sustainable forest management into the management of their forest resources. A few had also reported the development of timber certification schemes in partnership with appropriate stakeholders. Most European countries considered that their forests would now meet internationally accepted criteria for sustainable management. There seemed little justification now for any concern about unfair competition and double standards.

The consumer countries from the developed world had continued to provide technical and financial support to enable the developing countries of ITTO to enhance their capacity and capability in all fields related to sustainable forest management. On the whole, this was carefully targeted but, with a few exceptions, much of this assistance was not channelled through ITTO. Some of the reason for this was no doubt political (the preference of certain donors for certain recipients and for bilateral channels). But, many good projects designed to advance the Year 2000 Objective remained unfunded. The question arose: how seriously did the donors take the Year 2000 Objective?

During the period under review, all consumer countries reported a reduction or a phased reduction of import tariffs on timber and timber products, especially those from developing countries. Several countries, however, applied higher tariffs and duties for processed products. This was not helpful for those producer members who were striving to develop secondary and tertiary processing.

More could be done, in consumer countries also, by the trade, by professional foresters and by government to inform their public, and to correct misinformation, about the beneficial changes that were occurring in the management of natural tropical forests.

The report continued by discussing the ways in which ITTO had contributed to progress to date; this will be covered in Chapter 16. The recommendations in the report will be taken up in Chapter 15.

Reaction: false start, new energy

'The operative paragraphs ... were a masterpiece of the art of using oblique wording to satisfy all sensibilities; but they covered all important points. By such contrivances is international agreement reached'

The 28th Session of Council – Lima, May 2000

This Session started with a crowded agenda. On the international front, the UNFF had just been established and was faced with the challenge of implementing the 150 or so proposals for action that had emerged from the IPF/IFF. There were active discussions on the Kyoto Protocol of the UNFCCC and the possible implications of the Clean Development Mechanism – the implications for forestry were to be discussed in November. On the home front, things were beginning to move fast. The Council had a huge agenda before it; not only was there the Year 2000 Report to discuss, but business included auditing systems to assess sustainability, revision of the guidelines within a 'comprehensive framework', the extension of the Agreement, evaluating the success of projects, communication, certification, international activities, and relations with civil society – a formidable line-up for a one-week meeting. In the event, although much important business was completed,[1] the session got involved in interminable 'open-ended drafting sessions' in which valuable time was wasted in repetitive arguments over details of wording, many of which were of no importance for the operative parts of the Decisions. For the first time in Council history, a number of important draft decisions were deferred until the next Session – caused by a combination of a flood of important business, a failure to concentrate on essentials, and overbearing chairmanship. One almost incredible casualty was any Decision, or indeed any full discussion, on the Year 2000 Report. Indeed, the first Session in 2000 would have closed without any official recognition that the date of ITTO's landmark Objective had arrived, had not an eleventh hour appeal by the present author led to the agreement of a 'Statement':[2]

> ITTO affirms that it is fully committed to moving as rapidly as possible towards achieving exports of tropical timber and timber products from sustainably managed sources.

Mr William Mankin, who had been a constant and helpful participant in Council meetings, was impelled to intervene strongly.

> He had, during all these years, got to know ITTO as an organization that was very reluctant to do many things; to commit itself to the principle of sustainable forest management and, once it had made the commitment, to work really hard to achieve the objective; it had also been reluctant to provide credible evidence that progress was being made toward the objective; it took many years to acknowledge the existence of forest certification thereby not taking advantage of the opportunity it offered to its members; and it had been reluctant to embrace civil society representatives as working partners towards the goal of sustainable forest management.

Mr Mankin stressed that these aspects of the Organization were reasons why environmental NGOs, including the Global Forest Policy Project, had basically abandoned ITTO.[3] He went on to mention the many decisions that had been deferred, the weakening effect of that on the participation of members of civil society, and the absence of any decision on illegal harvesting. On the Year 2000 Objective, Mr Mankin regretted the failure to take any decision on helping to achieve rapid progress. 'The outside world would not take the statement of affirmation on the Year 2000 Objective as a strong commitment.'[4]

Some of these criticisms were well deserved, others overstated to the extent of being misleading. Did the fault lie with a weak commitment of Members and Council or with cumbersome and inefficient procedures? The environmental NGOs could equally be criticized for withdrawing participation.

The Chairman pointed out that 'the fact that Mr Mankin was allowed to participate in the drafting of decisions showed the openness of ITTO compared with other international organizations and NGOs, and that should be appreciated... Although there were no decisions on certain issues, it did not imply that those issues were irrelevant.' Rather, more time was needed for good decisions. He hoped that Mr Mankin would convey the right message about ITTO to the outside world.

The trade NGOs, on the other hand, were positive in clearly setting out their priorities based on their Market Discussions. These were: to provide statistical information of a high quality; to assist in value-added processing; to take a leading role in facilitating the exchange of information about national and regional certification schemes; to coordinate and facilitate the development of an effective market promotion strategy for tropical timber and timber products; and to continue to support RIL initiatives.[5]

Recommendations of the Year 2000 Report

As mentioned above, the Year 2000 Report and its recommendations were presented at this session. They were prefaced by this short statement on broad priorities, as perceived by the consultants, stressing once again the link between a flourishing timber industry and sustainable forest management:

Within the context of this review, three broad priorities have been identified. These are:

To develop a flourishing timber industry within every producer country while meeting all the criteria for sustainable forest management.

To disseminate information widely which demonstrates that well conducted and sustainable timber production is possible, and is compatible with all the conditions implied in the ITTO Criteria and Indicators.

To make ITTO into as efficient and effective an instrument as possible to assist producer countries in attaining the Year 2000 Objective by helping to mobilize the resources and experience of the consumer countries to this end and in minimizing any trade measures which might act in the opposite direction.

The recommendations that follow[6] are quoted directly from the Report in the present tense.[7]

(i)   Different countries are now at different stages in approaching the Year 2000 Objective; they also may meet different problems and are restricted by different constraints. These can best be diagnosed by discussion on the spot. It is recommended that *small, short diagnostic missions* (perhaps two people for two weeks) should visit a country, work together with the relevant ministries, departments, industries and other stakeholders, and visit a sample of forests, to decide where assistance can be most effectively and economically targeted.

(ii)   There is a lack of public knowledge and understanding of the progress that countries are making towards achieving the Year 2000 Objective. It is recommended: (a) first, that there should be a *clear and well publicized statement by ITTO of exactly what is meant by the Year 2000 Objective*, especially now that we are in the year 2000. There is, fairly understandably, a certain amount of confusion about this. (b) Secondly, that ITTO should *assist producer members in informing a wide public about their progress and disseminating information in an accessible form about these successes*, especially when they can demonstrate sustainable forest management on the ground. The public should also be made aware that sustainable forest management is not accomplished in a day, and that it will prove to be a hollow achievement (in fact not sustainable) unless it is built on solid foundations of local public support and a firm government legal and administrative backing.

(iii)   The next logical step after adopting and using criteria and indicators is the *establishment of standards of operation* which, in their turn, lead logically to certification of forest management; (timber certification can follow if there is a clear chain-of-custody). ITTO is the right international body to take the initiative in this field for tropical forests. There are a number of actions that might be taken:

(a)   ITTO to help the institutions in producer countries to *translate the ITTO criteria and indicators into practice* and, further, assist them in developing credible forest management and timber certification systems.

(b)   Engage in and become familiar with other schemes for criteria and indicators and for certification to *determine the extent to which they are compatible and could extend mutual recognition to each other.*

(c)   *Disseminate information on various certification schemes* operating in consumer countries, what they involve and how they are used. In particular, whether they establish conditions (either tariff or non-tariff) which might affect the import of timber or timber products.

(d)   *Help countries, on request, to develop their own criteria and indicators, and perhaps standards, compatible with the ITTO Criteria and Indicators.*

(e)   ITTO to be prepared to give a *stamp of approval to the criteria and indicators prepared by others.*

(iv)   Much more could be accomplished if countries which are advanced in the process of achieving the Year 2000 Objective were prepared to *share their experiences* with those which have not got so far. There is very great scope for *South–South co-operation*, perhaps also drawing in the North as an interested partner (and donor?).

(v)   ITTO should be closely involved in the planning and establishment of *demonstration forests*. These should be used for many purposes which would advance the Year 2000 Objective: to field test criteria and indicators and provide training in their use; to demonstrate how to proceed in protecting biological diversity within production forests; to provide training in the preparation of management plans, in RIL and many other aspects of sustainable forest management; to provide operational on-the-job training for visitors from other producer countries etc.

(vi)   It would be advantageous if countries were to exchange experiences (country to country) of the ways in which they had improved their performance and had overcome the difficulties in their way. It would be valuable if this could lead to mutual understanding of each other's procedures and regulations. If these could be harmonized, it might help in preventing the unscrupulous from taking advantage of the differences between them.

(vii)   In this respect, much could be done by networking between the leaders of ITTO projects within any region, and between ITTO projects and other relevant projects. This networking duty and possibly a duty associated with diagnosis [see (i) above] might be inserted in the terms of reference of project leaders.

(viii)   ITTO should encourage the formation of *partnerships between the private sectors of consumer and producer countries* to secure investment funds for sustainable forest management. ITTO could act as an ambassador and go-between in such negotiations.

(ix)   ITTO should encourage producer countries to involve *members of the timber industry* much more deeply in issues concerning sustainable forest management and should convince their senior managers of the importance of these issues. They could, for example, persuade concessionaires' professional associations to take responsibility for running training courses for tree fellers and extraction crews, or provide apprenticeships in various aspects of their trade directly concerned with sustainable forest management.

(x)   *Focus priorities within ITTO on issues connected to achieving the Year 2000 Objective and request consumer countries to align their official development assistance (ODA) accordingly.*

Two further overarching recommendations are presented separately.

(i)   Many of the above recommendations are concerned with catalytic actions, each of which should have a multiplier effect. They depend upon action initiated by the Executive Director and often carried out either by him or by the Secretariat staff. It is recommended that a *small fund should be set aside, to be used at the discretion of the Executive Director: (a) for catalytic actions such as those outlined in the previous set of recommendations; and (b) for very short missions (not more than two weeks) to deal with urgent problems arising in producer countries.*

(ii)   This Review has given an imperfect and incomplete account of progress. The time was too short; there was almost no possibility of field visits; and the format for the country reports was not well adapted to elicit the information required. It is recommended that any *future review should be quite different* and that its logistics should be carefully planned in advance. There are two possible alternatives, either of which should give adequate information:

(a)   *To rely on the ITTO Criteria and Indicators, which have been designed for the very purpose of recording change.* This would require: setting a date for the completion of the first return; and deciding an interval between returns. The return at the national level would have to be supplemented by evidence of sustainable forest management at the level of forest management units.

ITTO might be asked to be involved and should use this opportunity to train concessionaires in the completion of the return.

(b)   To conduct a *survey similar to that in 1988*, in which individuals experienced in sustainable forest management visited every ITTO country, examined the procedures and regulations in force to ensure that they covered the elements in the ITTO Guidelines and Criteria and Indicators, and conducted sample field checks to confirm the degree of compliance.

> It would seem that the time has come to reach a firm decision on the operational use of the ITTO Criteria and Indicators and on the way in which progress towards the achievement of the Year 2000 Objective can be authenticated.

Although the Report failed to lead to any formal Decisions in Lima, this was a failure of procedures rather than of will.

The 29th Session of Council – Yokohama, October–November 2000

The Session started with an up-beat address by the Executive Director. He set out a number of medium-term goals that he believed were achievable. These were:

- The establishment of 20 additional demonstration areas for sustainable forest management.
- The establishment of RIL training schools in each of the producing regions.
- The production of annual reports on the status of sustainable forest management at the national level applying the ITTO criteria and indicators.
- Concessionaires and other tropical timber producers to report their progress on the actual use of the *ITTO Manual for the Application of Criteria and Indicators* at the level of the forest management unit.
- The development of guidelines for the rehabilitation of degraded tropical forest land.
- The expansion of ITTO-sponsored transboundary conservation reserves to about 15 million hectares.

Other issues that required attention were illegal logging and illegal trade, which were undermining sustainable forest management and conservation in many countries. It was also important to provide training in the use of ITTO's criteria and indicators and thus once again take the initiative in this field.

Council was able in this session to gather momentum and to make up, to some extent, for its indecisiveness in Lima. The work programme began to take on a new and positive focus. Not all draft decisions could be finalized and some were again deferred to the next session, notably: on certification; illegal logging and trade in illegally harvested timber; auditing systems for criteria and indicators; and training in criteria and indicators and their use for reporting. There was also the new and important matter of preparing guidelines for the management of secondary forests. But, this time, the achievements were more significant than the deferrals. In particular, there was a strongly positive decision on the actions to be taken as a result of the Year 2000 Report – Decision 2(XXIX) ITTO Objective 2000. This, and Decision 3(XXIX) on Communication and Outreach Activities, accepted all the important recommendations in the Report. The operative paragraphs of Decision 2(XXIX) are reproduced in Box 15.1 – the term 'ITTO Objective 2000' was used to label the objective in its post-2000 form.

BOX 15.1 DECISION 2(XXIX): ITTO OBJECTIVE 2000

The International Tropical Timber Council ...

Decides to:

1. Reaffirm its full commitment to moving as rapidly as possible towards achieving exports of tropical timber and timber products from sustainably managed sources (ITTO Objective 2000);
2. Strongly encourage Member countries to use the ITTO Criteria and Indicators for reporting on progress made towards this objective;
3. Request the Executive Director to develop a format to facilitate the reporting of progress in implementing the ITTO Criteria and Indicators;
4. Invite Member countries to take concrete measures to enhance market access for tropical timber from sustainably managed sources;
5. Authorize the Executive Director to render assistance to producer countries, on request, to identify in each country those factors which most severely limit progress towards achieving Objective 2000 and sustainable forest management and to formulate an action plan to overcome these constraints...;
6. Facilitate the implementation of these action plans through intensifying international collaboration and strengthening national policies and programmes by measures such as demonstration projects and reduced impact logging training facilities;
7. Authorize the Executive Director to assist individual producer countries, on request, in setting up an ITTO Objective 2000 Board or nationally appropriate focal group to build both broad-based support and high level commitment in:
   - providing a focus for efforts in achieving Objective 2000; and
   - marshalling internal resources so as to ensure their optimal application in achieving Objective 2000;
8. Authorize the Executive Director to seek voluntary contributions from Member countries to meet the financial requirements of paragraphs 3 and 5 of this decision. If sufficient contributions are not received by 31 December 2000, the Executive Director is requested to use funds from Sub-Account B of the Bali Partnership Fund;
9. Urge donor Member countries to make increased financial contributions to the Bali Partnership Fund to facilitate the attainment of Objective 2000 and sustainable forest management in producing countries; and
10. Authorize the Executive Director to seek earmarked funds to meet the financial requirements of paragraph 7 of this Decision.

Source: ITTC(XXIX)/31.

Decision 2(XXIX) broke new ground in the discretion given to the Executive Director – for example, it allowed him to spend up to US$100,000 each for as many as five producer countries per year to help such countries identify those factors most severely limiting progress towards achieving Objective 2000 and sustainable forest management, and to formulate action plans to overcome such constraints. The Bali Partnership Fund was at last starting to fulfil the intention of the 'reforestation fund' envisaged by the founders of ITTO.

Other positive decisions dealt with a new ITTO Action Plan, mangrove conservation and the international role of ITTO. Moreover – in a move halfway between the full missions of the past and the new 'diagnostic missions' –

Indonesia requested a technical mission 'to assist the Government of Indonesia to identify ITTO support, especially in formulating forestry action plans to achieve sustainable forest management in Indonesia'.[8]

Members had been upset by the vacillation shown at the last meeting and there was outspoken discontent about the deficiencies of the Organization. Mr Rob Rawson of Australia considered that:

> the session was characterized by considerable frustration for many Members and Observers, particularly because processes did not seem to support the broader interests of members in relation to decision making... All was not well with ITTO and there was a need for changes to be made as quickly as possible... Changes should seek to build greater trust and transparency as well as de-emphasize personalities and the pursuit of narrow interests... Australia recognized the value of ITTO projects and activities which were delivering worthwhile outcomes for Members, although there was a perception that progress towards sustainable forest management had been slow and that the Organization was not committed to tackling the more difficult issues within its mandate... If ITTO could not make decisions on the politically sensitive issues such as certification and illegal logging and illegal trade, some Members might look for other ways to cooperate and work together without ITTO... He would like to see these issues on the agenda of the next Session.

Although expressing equal concern, Ms Jan McAlpine (US) had confidence that ITTO could rise above its difficulties. She expressed concern:

> about [the] frustration of the outside world with the ability of ITTO to effectively address key issues within the scope of the ITTA... In the last few years the Organization had not been able to figure out a process for effectively considering substantive issues... Issues like illegal logging and certification [were] sensitive issues that should be addressed by the Council... In spite of the problems, there was a true commitment on the part of all Members... ITTO [had] the potential and the foundation for making immense contributions to the Objective 2000 and all the issues in the ITTA, 1994.

The Yokohama Action Plan 2002–2006

The Yokohama Action Plan 2002–2006 was adopted in November, replacing the Libreville Action Plan which was designed to run until 2001.[9] In preparation for its successor, a study was commissioned, and an Expert Panel set up, to look into the form of the new action plan.[10] The Working Paper for the Panel did not prepare a draft action plan; rather it provided 'a detailed analysis of interpretation of past activities'. In particular, it examined the work of ITTO between 1998 and 2001 and compared it with the proposals made at Libreville.

The working paper began by reviewing changes since Libreville. Although the review of the scope of the ITTA, 1994 had not yet taken place, ITTO now had certain responsibilities in relation to non-tropical timber and non-tropical forest issues; the year 2000 had now passed and progress had been reviewed; the Bali Partnership Fund was now operational; there was a need to review the seven priority areas previously agreed; ITTO now had new criteria and indicators, and training was being provided in their use with the aid of the 'format'. The consultants continued:

> Notwithstanding the importance of securing an agreed framework for SFM, it is considered than an adequate framework is now available and that it would now be desirable to concentrate more effort on improving the field implementation of SFM... The process has been well established but the hard actions have been less noticeably addressed.
>
> It seems that neither the adoption of the Libreville Action Plan nor the operationalization of the Bali Partnership Fund has, so far, significantly increased the investment made by the Organization in support of its Members in actually improving management on the ground.

They argued, moreover, that sustainable management of the resource was only part of the picture; success also depended upon industry and the market. Sustainable management incurred increased costs, prices would have to rise and competitiveness would be prejudiced. 'There [would have] to be major changes in efficiency along the whole supply chain and innovative approaches to secure market share.' Since Libreville, there had been a downturn in the market. Certification and other mechanisms would help to avoid market barriers and impediments, but they would not be effective in increasing either demand or prices. If the market was to be the main source of finance, market access had to be assured. The competitiveness of tropical timber products needed to be examined vis à vis other timbers and other competitive products. It followed that more attention should be given to forest industry and market issues. In the past, there had been a great preponderance of funding devoted to the resource.[11]

The conclusion of the consultants was one of very considerable significance:

> The full benefit of such investment [in the resource] requires equivalent commitment to the development of efficient forest industries and sound marketing based on good information [in order] to secure a profitable sector that will generate funds for reinvestment in sustainable forest management. This will need to be reflected in defining the priorities, goals and action in the revised Action Plan.

This change in balance was reflected in the report of the Expert Panel and in the Action Plan finally agreed by the ITTC – the *ITTO Yokohama Action Plan 2002–2006*. This follows much the same arrangement as its predecessor, but has some important differences.

BOX 15.2 YOKOHAMA ACTION PLAN: KEY STRATEGIES FOR IMPLEMENTATION

One of the major focuses for ITTO and its members during the term of this Action Plan is accelerating progress towards the fulfilment of Objective 2000, ie moving as rapidly as possible towards achieving exports of tropical timber and timber products from sustainably managed sources. This requires:

1. An integrated approach across all three areas of ITTO's work, balanced as appropriate: *Economic Information and Market Intelligence, Reforestation and Forest Management* and *Forest Industry*. Examples of such integration include: increasing efficiency and value-added throughout the production chain and further development of markets for tropical timber and timber products from sustainable sources.
2. Shifting focus from development of national forest policies and legislation toward the implementation on the ground, especially at the forest management unit level. This would include, for example: supporting stronger forest law enforcement; greater training and capacity building; wider application of reduced impact logging; and strengthening timber tracking to improve the accuracy and transparency of information on timber products and trade. This shift of focus from policy development to implementation is also applicable to forest industry and marketing.
3. Filling key knowledge gaps. These could include, for example: assessing sustained yield capacity; consumer preferences; competitiveness with non-tropical timber products and timber substitutes; and better understanding of the social dimension of sustainable forest management and of the supply chain.
4. Diversifying incentives for maintaining and expanding the forest base to help assure continued timber supplies. This would include factoring in the value of, and developing innovative markets for, ecosystem services derived from production forests.

Source: ITTC(XXXI)/26.

It selects four key strategies for implementation. These (see Box 15.2) represent a firm shift in balance.

The Action Plan also proposed some 'action for effective implementation'. It will be evident from earlier chapters that the way in which ITTO has operated is far from efficient in delivering its objectives. The proposals in this Action Plan – if implemented – would go far to remedy this deficiency. Some of the most important proposals are: the reallocation of responsibilities between Council, Committees and Secretariat; organizational changes to Council sessions and improved preparation for Council decisions; greater latitude to the Secretariat for 'proactive actions'; tightened links between the annual work plan, the Action Plan and emerging priorities; closer correlation between project submissions and ITTO's objectives; improved monitoring of projects and communication of lessons learned; greater participation by members in diagnostic missions, training and statistical reporting; the encouragement of projects on industry and market issues; and the examination of ways to broaden

ITTO's funding base. Finally, there should be a mid-term evaluation of the implementation of the Action Plan.

In general, the recommended actions are much more realistic and tightly defined than in previous versions. Whether or not they are delivered effectively will greatly depend upon the resolution of the Council to reform its procedures. Moreover, since the Action Plan does not offer any prioritization for the recommended actions, the donors presumably will continue to set *de facto* priorities by predominantly financing those projects that conform to their own national objectives.

The Mission to Indonesia and Illegal Logging

In 2000 the Minister of Agriculture and Forestry of the Republic of Indonesia requested a technical mission to identify ITTO support, especially in formulating plans and programmes to achieve sustainable forest management. This was the first major mission since that to Bolivia.

The report[12] is most discouraging and very much at variance with the country report submitted by Indonesia for the Year 2000 Objective.[13] The initial abstract sets the scene:

> The current state of forests is increasingly disheartening. The most glaring aspect of the present situation of the Indonesian forest sector is the unsustainable management of forests under the current system of forest land management. This in turn has led to deforestation, environmental degradation, rampant illegal logging, increasing incidence of forest fires, falling productivity and widespread land conflicts.

The situation is clearly very serious indeed; it can genuinely be called critical. Eighty per cent of Indonesia's primary forests had been logged; there was much uncertainty about the productivity of the second growth forests and, therefore, about future economic supplies of timber. There is no evidence in the report that any of Indonesia's 'production forests' are under sustainable management.

The report is frank and the recommended reforms are numerous and far reaching – far too many to be covered in this chapter. Under its terms of reference the Mission made recommendations on five issues of particular importance: curbing illegal logging; restructuring the forest industries; forest plantations for resource creation; recalculating timber values; and implementing decentralization of the forestry sector.

One of these issues, illegal logging, has been selected for this account to illustrate the extent of the problem and the recommended remedies. This has been chosen because it is such a topical issue and because, in Indonesia, it casts a deep shadow over all other issues. To quote:

> Illegal logging is now recognized as one of the most critical problems of forestry and forest industry in Indonesia. Reports indicate that illegal logging has far exceeded the legal production. Estimates ... vary

considerably and range from 25 to 57 million m³ annually, from 52 percent to 70 percent of total log production... Some claim that cross-border smuggling alone may account for about 10 million m³. In 1998, illegal log production, in a total production of 77 to 79 million m³, amounted to an estimated 57 million m³, while the AAC for Indonesia was of the order of 20 to 22 million m³.

The World Bank had estimated that, based on 1980 tax rates, if all timber taxes had been collected, the government would have received an extra US$1.2 billion over the period 1980–1985. The amount seems to have grown to about US$2 billion per annum in 1998/1999 without considering the losses in terms of ecological costs. Some estimates place the total annual loss to the country from illegal logging at about US$3.5 billion...

Some reports indicate that, of the remaining natural forests in Indonesia, 38 percent of national parks and conservation areas, 46 percent of protection forests and 30 percent of production forests have already been degraded... Forest plunder in Indonesia seems to have particularly affected the integrity of national parks and the protected area system.

The report analyses the direct causes of illegal logging as: an unmet demand for logs; the high profitability of illegal logging operations; the 'greed revolution'; weak law enforcement; the availability of a ready market abroad for illegal logs; and unrealistic assumptions of conversion rates. Indirect causes are: the low risk of illegal logging operations; rural poverty and unemployment; conflicting land use policies; lack of coordination with other sectors; and local tenurial anomalies. A number of 'actors and accomplices' are involved. Among the actors are: illegal log buyers, concession holders, investors looking for a quick profit, businessmen who collude in cross-border trade and export of illegal logs, and unscrupulous elements in enforcement agencies. Minor accomplices include: poor and unemployed people looking for some income; disadvantaged and disenfranchised tribal communities; jobless and disillusioned youth; local community leaders; transport agents; and distributors of illegal logs. In fact, the problems lie deep in the social and economic fabric of the country.

The Mission's proposals for action on illegal logging are given in Box 15.3.

The report suggests project ideas to support this programme. These should be pilot projects to prepare the way for wider adoption. The training of trainers in the use of criteria and indicators is already part of an ITTO project. Other proposals are the 'development and implementation of guidelines to control illegal logging in Riau and West Kalimantan' (an ITTO project is under way at the time of writing) and 'the application of forest cover monitoring – satellite imagery to verify illegal logging'. The proposals illustrate the very weak leverage that can be exerted by projects in such a situation.

Over the whole field that it examined, the Mission has 47 recommendations for action divided into six main heads, five of them the issues identified above, and the sixth concerned with institutional, legal and financial prerequisites. From these, six were proposed for immediate implementation, summarized below:

BOX 15.3 ACTION PROPOSALS TO CURB ILLEGAL LOGGING

Principles

- Carry out law enforcement without fear or favour, increasing the level of deterrence wherever appropriate.
- Strengthen the forest sectoral institutions to be capable of handling forest offences and other illegal activities.

Short and medium term actions

- Undertake necessary policy and institutional change to facilitate law enforcement on the one hand and to provide incentives to those who behave lawfully on the other – eg temporary log export ban.
- Establish an anti-logging task force answerable directly to a high authority (preferably the President or Vice-President of the Republic) and attack illegal logging on a war footing.
- Establish mobile squads with appropriate powers to apprehend illegal loggers.
- Restore tenure rights of local communities over adequate extent of land, through clear policy pronouncements.
- Regulate capacity of wood industries; introduce a system of vigilance and inspections; strictly enforce transparent timber accounting and log audit for all wood-based industrial units.

Long-term actions

- Modify and restructure the 'Forest Concession Rights' (HPH) System such that it will be in the interests of concession holders to cooperate in countering illegal activities.
- Use the leverage of the Indonesian Ecolabelling Institute (LEI) in promoting sustainable forest management and to curb illegal activities; introduce a system of rewards to encourage such involvement.
- Conduct an in-depth analysis of the root causes of illegal logging.
- Bestow attention to socio-economic welfare of the local community and involve them as partners in fighting illegal logging.
- Also, test other systems of forest management found successful in other countries and, if found feasible, introduce them in Indonesia to reduce illegal logging.
- Address the issue of corruption within enforcement agencies.

Source: ITTC(XXXI)/10.

- The establishment of a National Forestry Council, the Ministry of Forestry to provide the secretariat and task forces (eg anti-illegal logging), and to formulate the National Forest Programme according to IPF/IFF guidelines.
- The strengthening of policy and law enforcement to curtail illegal logging – eg policies in relation to settling tenurial issues in forest lands.
- A strategic plan for restructuring forest industries.
- A phased strategic plan for decentralizing sustainable forest management, involving levels of decentralization, division of functions and responsibilities and accountability, but avoiding over-regulation.
- Rationalization and streamlining of Industrial Timber Estates with special emphasis on joint ventures.

- The preparation and implementation (with international assistance) of a project to demonstrate models of decentralized and multi-stakeholder forest management in selected districts.

The Mission report was received by the Council with much interest and appreciation. It led to Decision 5(XXXI), 'Strengthening sustainable forest management and controlling illegal logging in Indonesia', with operative paragraphs as follows:

> 1 Encourage Indonesia to submit project proposals implementing the recommendations of the Technical Mission, in particular those seeking to curb illegal logging, restructure forest industries, recalculate timber values, enhance the development of forest plantations for resource creation, and decentralize the forestry sector; and
>
> 2 Disseminate the findings of the Technical Mission through a workshop for national, provincial and district officials of relevant ministries and departments.'

Forest Law Enforcement

Illegal logging and trade in illegally logged timber had long been a delicate matter in ITTO, for many producer countries considered it to be purely an internal concern. But pressure had been growing for some time for ITTO to act; at the 29th Session the Executive Director had reported to Council that there was ample evidence that sustainable forest management and forest conservation efforts were being undermined by illegal logging and illegal trade in many countries; they were barriers to further progress towards the ITTO Objective 2000.

By the 31st Session (Yokohama, November 2001) the time was ripe to take matters further. The issue was first raised by the Minister of Forestry Economy and Fisheries of the Republic of Congo (also at the time the Chairman of the African Timber Organization Ministerial Conference). He expressed concern about illegal logging; the need for rational utilization of natural resources was beyond the narrow framework of countries; it required global approaches. Mr Enzo Barattini of the EU emphasized that there was now a stronger commitment, evidenced by the Ministerial Declaration issued at the end of the East Asia Ministerial Conference on Forest Law Enforcement and Governance, which had been held in Indonesia in September. The topic was also picked out by the Executive Director in his opening remarks. He urged members to cooperate to protect the forest estate from illegal activities, without losing sight of the fact that the prevention of illegal logging and illegal trade were primarily the responsibility of national and local authorities. He proposed that ITTO could focus on providing assistance to these authorities, mostly in the form of information and data, situation analysis and recommendations, and putting in place appropriate policies and measures to prevent these transgressions. The Executive Director further proposed that Council might, on request, finance case studies of illegal logging and illegal trade in Member Countries; it might also consider formulating ITTO guidelines on the subject. ITTO must ensure that the issue of illegal logging and illegal trade did not become a pretext for protecting local industries in importing markets by restricting the access of tropical timber. These matters were also raised in the context of CITES Appendix III listings and of inconsistencies in trade statistics. Colombia, for example, attached great importance to CITES as a way of preventing illegal trade. In the Council debate, both the urgency of this issue and its delicacy were reiterated; some of the difficulties might be resolved by a change in terminology.

Terminology, in fact, proved to be part of the problem; illegal logging and trade in illegally harvested timber was the subject of a firm Council Decision[14] under the title 'Forest Law Enforcement in the Context of Sustainable Timber Production and Trade'. The operative paragraphs, reproduced below, were a masterpiece of the art of using oblique wording to satisfy all sensibilities; but they covered all important points. By such contrivances is international agreement reached.

> The ITTC decides to:
>
> 1 Authorize the Executive Director to engage consultants to conduct, with producer and consumer countries interested in voluntarily participating, a case study on assessing export and import data on tropical timber and timber products in the context of international trade, with a view to improving the accuracy of ITTO's market and economic intelligence;
>
> 2 Encourage countries in need of ITTO assistance to voluntarily submit projects for the Organization's consideration which address unsustainable timber harvesting, forest law enforcement and illegal trade in tropical timber with a view to attracting increased funding from Members to develop domestic capacity to address these areas as a matter of priority;
>
> 3 Consider in future, in cooperation with other relevant international organizations, the implementation of a global study to assess the extent, nature and causes of illegal trade in timber and timber products and to make recommendations on its prevention in the context of the ITTA, 1994;
>
> 4 Request the Executive Director to compile and analyse information on relevant issues affecting market access for tropical timber and to present this to the Thirty-third Session for the Council's consideration and action as appropriate;
>
> 5 Authorize the Executive Director, upon request by producer countries, to engage consultants to conduct studies and to assist Producer Countries in devising ways to enhance forest law enforcement, taking into account, when necessary, illegal timber trade and its impacts;
>
> 6 Invite Members to include in their reports on Progress towards Objective 2000 information on the promotion of sustainable forest management and timber production and trade practices, including inter

alia forest law enforcement as well as market access obstacles to tropical timber and timber products, and to share this information among Members;

7 Request the Executive Director to follow the work of international organizations in this area and to contribute as appropriate and to report to the Council; ...

The International Tropical Timber Organization in 2001

The 30th Session was held in May 2001 in Yaounde and the 31st in November, in Yokohama.

Certification was once again on the agenda – and finally reached the stage of a Council Decision.[15] The Executive Director, again:

> it was a matter of some concern that, while the emergence of certification in the early to mid 1990s was due, in large part, to the clamour for improvements in tropical forest management, its implementation had been largely occurring outside the tropics... Recent data indicated that 18.3 million ha of the 21.7 million ha of forest certified by the Forest Stewardship Council (FSC) were located in Europe, North America and New Zealand. Of the remaining 3.4 million ha, only 1.8 million were [in] ITTO producing Member countries.[16]

In fact, as far as ITTO was concerned, certification had had the perverse effect of improving market access for competing timbers. The Trade Advisory Group wished 'ITTO to be actively involved in facilitating moves towards the mutual recognition of timber certification schemes and not endorse or be perceived to endorse any one particular scheme'.

Another decision provided funds for ten national training workshops in the use of the format to report on criteria and indicators, a major step towards translating international policy into field practice; and strong encouragement to producers to submit their first national level report by the end of 2001. These should use the format and include a 'summary of highlights'. They would contribute towards a 'Status of Tropical Forest Management Report' to be prepared and published by ITTO. At last, perhaps, it would be possible to assert with confidence how much tropical forest was under sustainable management for timber production.

There was much international activity during the year. ITTO was an active partner in the CPF. It was involved in the planning of a joint meeting on criteria and indicators with FAO. Cooperation with IUCN was being strengthened. There was contact with UNDP and the Security Council on 'conflict timber' (timber being sold to acquire arms) – prevalent in the Democratic Republic of Congo, Liberia and Cambodia. There was work on a paper for the Global Environment Fund on biodiversity and certification.

ITTO was now more closely involved with CITES; it was regularly consulted about proposals for listing, was taking part in the revision of the criteria for Appendices I and II, and in a study of the effectiveness of Appendix III listing in relation to mahogany. The initial mistrust had disappeared in a new practical working relationship; it was becoming recognized that inclusion of a tree species in Appendix III could be advantageous in controlling the movement across frontiers of illegally harvested timber, for example of ramin in Indonesia and of mahogany and *Cedrela odorata* in Colombia and Peru. It was also evident that transboundary protected areas, as well as their special importance in conserving biological diversity *in situ*, were to be a significant additional deterrent to illegal trade in timber and endangered plants and animals. ITTO played an important part in convincing the Cameroon Government to double the area of the Mengamé Gorilla Sanctuary from 65,000 to 122,000ha and in gaining the commitment of Gabon to cooperate in the establishment of the Mengamé-Minkebé Transboundary Gorilla Sanctuary.

The importance of missions was being recognized once again. Apart from the Mission to Indonesia, requests for diagnostic missions had been received from Brazil, Cameroon, Congo and the Central African Republic; and requests for the setting up of Objective 2000 Boards had been received from Cameroon, Congo, Gabon and Papua New Guinea. The aims of these were to provide a focus for efforts in achieving Objective 2000 and to marshal internal resources to ensure that they were effectively applied.

Many of these significant activities would not have been possible without the Bali Partnership Fund. At the 26th Session, a Sub-Account B of the Fund had been established 'for actions captured under the goals set by the current ITTO Action Plan that [were] non-country specific'. This gave ITTO considerable scope to implement those projects and activities deemed to have application at the regional or trans-regional levels and has thus increased its freedom of action. It now can move swiftly to apply closer to the ground the policies arising from Council decisions. Thus, by the 32nd Session, 34 projects, pre-projects and activities with a total budget of US$8.9 million had been funded from this source. The main donors were Japan, Switzerland and the US; also, 50 per cent of the interest earned on the Special Account is contributed to Sub-account B. Among other activities funded were the short missions, the action plan, the manual and format for criteria and indicators, communication and outreach programme, cooperation with the ATO and the study on auditing systems.

After the frustrations of 2000, the Council had entered 2001 in a very positive mood and, by the end of the 31st Session, had good reason to be satisfied with its achievements. It had agreed a new and improved action plan and reached decisions on controversial issues such as certification, forest law enforcement, and auditing systems for criteria and indicators. Discussion had begun on strategic planning to make ITTO into a more effective instrument for carrying out its mission. Delegates were very happy. The 'delegation of Switzerland considered the Thirty-first Session a great success'; the EU 'expressed satisfaction at the decisions taken'; Ms Jan McAlpine (US) said that the 31st Session represented a very important chapter in the history of ITTO.

Policies in action: some case studies

'ITTO has the advantage of having both the mechanism and available funds.
Although these fall far short of what is needed, they do allow countries to make
a start and to build capacity through a process of "learning by doing"'

An introduction to the working of the ITTO project system was given in Chapter 4, but there has been little mention of projects in subsequent chapters. Nevertheless, projects have been formulated, selected and funded during the whole life of ITTO. Indeed, Decision 1 at each session of the Council is devoted to projects. This chapter concentrates on the contribution that ITTO projects have made to sustainable forest management and a sustainable timber trade. It is supplemented by four case studies.

One of the main incentives for producer countries to belong to ITTO has been the possibility of obtaining funding through projects. Projects are crucial to the work of the Organization, for they are an important means by which policies are realized in producer countries and in the forest. As the policy debate matures, this linkage between international policy development and field action becomes crucial. Policies developed in the stratospheric atmosphere of international fora are useless if they are never implemented; some might argue that they are worse than useless, because they give the impression of activity when in fact there are no changes where they really matter. The UNFF was created to realize the IPF/IFF proposals for action. It seems that the UNFF debate has shifted somewhat recently from what needs to be done to how it is to be done. Yet it is still a debate, not action. Indeed, some environmental NGOs have recently become sceptical of the UNFF, partly because it has no mechanism for implementing its many proposals. ITTO has the advantage of having both the mechanism and available funds. Although these fall far short of what is needed, they do allow countries to make a start and to build capacity through a process of 'learning by doing'.

The focus, quality and effectiveness of projects should therefore be significant measures of ITTO's success. In all these respects the record is mixed. Much attention has been given since 1987 to improving quality through the progressive refining of procedures for the preparation, review and approval of projects. Training courses have been given in the preparation of projects, a project manual has been prepared and projects are rigorously screened and reviewed; yet many proposals still have to go through several revisions before they are considered satisfactory.

Also, although some priorities have been set in the successive ITTO action plans and in the Year 2000 Objective, the match between projects and priorities has often left much to be desired. It must be admitted, however, that the action plans have tended to be all-inclusive shopping lists and there has been uncertainty about the exact scope of the Year 2000 Objective. In fact, the circumstances of the producer countries are so different from one another that it may be counter-productive to attempt to prepare a list of general priorities. The differences are illustrated by Table 16.1.

It has not been easy, either, to decide where assistance should best be targeted. Producer countries have been surprisingly reluctant to report on their difficulties in either the Mid-Term Review or the Year 2000 survey. This should improve as more countries are visited by diagnostic missions or submit reports based upon criteria and indicators. Meanwhile, much has been accomplished by the Secretariat, who have discussed priorities with producer countries and assisted in the formulation of suitable projects. It is hoped that these processes will be made easier and swifter by the new wider discretion given to the Executive Director. Another serious past deficiency has been the lack of any ex post evaluation; but this is now being remedied.

Unfortunately, many good projects have been approved but have failed to attract funds. Until the setting up of the Bali Partnership Fund, all money for projects had to come from voluntary contributions from consumer countries – the vast majority from Japan (75 per cent), Switzerland and the US. The total was wholly inadequate to fund even those projects approved by Council. Donors have been very selective, looking for high-quality projects which fit both ITTO's priorities and their own. Also, any voluntary contributions made to ITTO reduced the amounts that they could use in their own bilateral aid programmes which they might feel were more likely to be effective. More cynically, through bilateral aid the donors can provide employment for their own nationals and ensure that many of the goods and services purchased for projects come from their own countries. Some donors have explicit policies that at least 50 per cent of official development assistance (ODA) is spent in this way; others may not state the policy explicitly but the end result may be similar.

ITTO's project programme is quite different from most bilateral programmes and those of many multilateral institutions. In ITTO, projects are country-driven. Project proposals are developed by the recipient country and implemented by agencies nominated by that country – governmental, non-governmental or both. ITTO's role is to scrutinize and approve project proposals, provide funding and oversee and monitor implementation. Projects thus have a high level of in-country ownership, provide much local employment and bring other incidental local advantages. Countries have the opportunity to set their own priorities, decide how to put these into practice and seek funds from an international agency to move forward.

Table 16.1 *Diversity among Producer Members*

Significant forest industry sector		Small or relatively undeveloped forest industry sector		Net consumer
Large resource base	*Limited resource base*	*Large resource base*	*Limited resource base*	
GNP/cap < US$695				
Cameroon	Côte d'Ivoire	Cambodia	Honduras	India
	Ghana	CAR	Liberia	Togo
		Congo		
		DR Congo		
		Guyana		
		Myanmar		
GNP/cap US$696 to 2785				
Indonesia	Ecuador	Bolivia		Panama
		Colombia		Philippines
		Fiji		Thailand
		Peru		
		PNG		
		Surinam		
		Vanuatu		
GNP/cap US$2786 to 8625				
Brazil				Trinidad
Gabon				Venezuela
Malaysia				

Source: ITTC(XXXI)/7.

In theory, at least, this seems a very effective way in which developing countries can make use of the sparse funds offered through ODA. It carries the danger, however, that accountability is low; in its early days, ITTO was seen as a 'soft touch' for project financing. Since then, the requirements for financial accountability have been significantly raised[1] and institutions that do not comply receive no disbursement of funds until these requirements are met. ITTO is now, at last, evaluating the success of past projects in order to improve future projects. Where countries have limited capacity to formulate and implement sound projects, ITTO helps them to become more competent in these fields.

In spite of these difficulties and limitations, most of which stem from the original rejection of the proposal to establish a Reforestation Fund, ITTO projects can chalk up some considerable successes. Since 1987, over 400 have been funded with a total value of US$220 million – many of them relatively rapid responses to ITTO's own policy work and to Council decisions. There is inevitably some time lag between the formulation of any new policy and its implementation; but ITTO can often ensure that this is reduced to a minimum by working with countries to develop relevant projects which it can then fund immediately from its own resources. This is one of its strengths.

Policies and Projects

Earlier chapters have given some insight into the way in which policies and priorities are shaped by the work of the Secretariat, the studies of consultants, missions, expert panels and working groups, and, finally, by the debates, discussions and decisions of the ITTC. We have seen, for example, how the ITTO/IIED study (a pre-project) pointed the way to many policy initiatives – the action plans, the guidelines for natural tropical forests, for plantations and for biological diversity, and eventually to the general form taken by the Mission to Sarawak and the criteria and indicators. These all had a significant effect in the producer countries on the development of attitudes towards sustainable forest management and on the evolution of forest policies. In some, such as criteria and indicators, ITTO blazed a trail that was followed by the rest of the world; in others, such as the Mission to Sarawak, ITTO pioneered an approach that proved unusually influential. The same point can be illustrated by the other policy initiatives that have been taken by ITTO and described in previous chapters, such as the HIID and LEEC studies and reduced impact logging.

As time went on, the climate of opinion arising from ITTO activities had significant consequences on the formulation, acceptance and funding of projects; ITTO producers more and more tended to submit proposals, and consumers were more ready to provide funds, for projects that conformed to the identified priorities. This can be illustrated by the sequence of events leading from the guidelines to the preparation of national reports.

The guidelines – and the criteria and indicators – encouraged member countries to request project assistance to put the theory into practice. For example, the project 'Development of national guidelines for the sustainable management of natural tropical forests in Cameroon' was designed to translate the ITTO guidelines into national guidelines adapted to the forest policy of Cameroon. In Brazil, one project first issued 4500 copies of the two publications translated into Portuguese, and then over the next three years promoted their dissemination and training in their use. According to the final report, this project has resulted not only in a greater awareness of ITTO's work and of these policies, but to a complete revision of the field application of existing management plan regulations; support for activities related to the preparation of regional and national guidelines and criteria; the production of a data bank on forest management plans for the State of Para; and recognition by the local timber industry in Para of the importance of planning as a means of reducing environmental impacts and ensuring sustainability. More recently, ITTO has funded projects in both Cameroon and Gabon where the revised criteria and indicators have been tested to adapt them to national conditions.

There have recently been two reviews of the relation between projects and priorities – the first as part of the Year 2000 Report (looking at the match between projects and the Year 2000 Objective) and the other in preparation for the Yokohama Action Plan (comparing the projects funded with the recommended actions in the Libreville Action Plan).[2]

Table 16.2 *Distribution of projects and funding according to subject area and category*

Subject area	Number of pre/projects			Funding (US$ million)			Percentage of total	
	I	*II*	*Total*	*I*	*II*	*Total*	*Number*	*Funding*
(a)	13	1	14	4.51	0.60	5.11	4.8	3.7
(b)	27	38	65	18.77	19.19	37.96	22.6	27.4
(c)	8	9	17	2.98	2.92	5.90	5.9	4.3
(d)	13	8	21	37.00	4.74	41.74	24.7	30.1
(e)	51	20	71	5.79	2.29	8.09	7.3	5.8
(f)	7	7	14	6.82	2.30	9.11	4.8	6.6
(g)	23	6	29	12.89	2.74	15.63	10.1	11.3
(h)	12	45	57	3.92	11.02	14.94	19.8	10.8
Total	154	134	288	92.68	45.80	138.48	100.0	100.0

Source: Poore and Thang, 2000.

Projects and the Year 2000 Objective

By the end of 1999, many projects and pre-projects (scoping or feasibility studies) had been or were being carried out by ITTO in support of the Year 2000 Objective. At that time, they numbered 288 with a total budget of US$138.5 million – a significant amount to have invested in sustainable forest management and the Year 2000 Objective. These projects and the funds provided are summarized in Table 16.2. They are divided into two categories (I – having a direct effect, and II – having an indirect effect) and eight subject areas based broadly on the ITTO criteria, as follows: (a) policy and legal issues; (b) institutional strengthening; (c) forest resource; (d) forest health; (e) sustainable forest management; (f) conservation; (g) community participation; and (h) trade and industry. A short description of the first seven follows; the relative weakness of projects in trade and industry will be discussed later.

Policy and legal issues

These are projects where ITTO has supported the formulation, updating and clarification of national forestry laws and regulations, and of guidelines and criteria and indicators for application to national forests. In Bolivia, for example, there was a 'Forestry law regulation' project, one of the outputs of which was technical standards and regulations for the classification of land capability and forests. In Peru, a draft law was developed to establish the legal basis for sustainable forest development. In Cambodia, a project aimed to establish a strategy for sustainable forest management as a precursor to a longer-term master plan. In Colombia, the national forest situation has been reviewed in order to establish a national policy for the forestry sector. The projects that followed the Missions to Sarawak and Bolivia have already been described.

Institutional strengthening

ITTO has contributed US$5.6 million towards the long-term strengthening of local capacity in forest resource management in Indonesia through a three-phase project. This has involved improvements in forest research, in the management of conservation and protection forests, in human resource development and in public awareness. In Brazil, one project is aimed, amongst other things, at strengthening the ability of the sector to improve the management and use of renewable resources. A project aimed at all ITTO producer members works with local institutes to develop curricula for the planning and management of forest industries and to develop their capacity to continue to offer such training courses. ITTO has provided consistent support in building up a national capacity in forestry statistics (see case study 4 below).

Since the ITTO Fellowship programme began in 1989, up to mid-2002, 680 fellowships, totalling about US$3.9 million, had been awarded. The fellowships have provided individuals with the opportunity to pursue studies, attend seminars or training sessions or carry out research. The programme has been revised to take account of later priorities. More indirect but important contributions to human resource development were 14 workshops held in different member countries under a project devoted to the identification and formulation of projects.

The forest resource

A group of projects has focused on establishing and/or securing the forest resource. Examples are: 'Development of a master plan for forest management in Congo', 'An assessment of the status of conservation areas in Sabah, Malaysia', and 'Integrated buffer zone development for sustainable management of tropical forest resources in Thailand'. Also included were some projects on the establishment of plantations. Although gaining financial support for projects involving plantations has generally proved less easy, ITTO has encouraged some work in this area, for example for a 2500ha timber production plantation in Togo and the development of teak cloning and industrial teak plantations in Côte d'Ivoire. In Colombia, a project directed at institutional strengthening for the sustainable management of forest plantations has helped to focus the attention of investors on plantations and has also raised the issue of carbon sequestration.

One of the obstacles to sustainable forest management identified by the 1988 ITTO/IIED report was the problem of encroachment and illegal logging. This was a subject that ITTO had not really confronted by 2000. Further developments have been described in Chapter 15.

Forest health

Many of these projects, which address problems related to the condition of the forest ecosystem, come in response to the need for forest rehabilitation and reforestation in areas where land has been badly degraded. Two examples of such projects are 'The recovery of natural systems of the hillsides of Caqueta' and 'Pilot project for the reforestation and rehabilitation of degraded forest lands in Ecuador'. Such projects also include the establishment of nurseries and production of propagules. Where the projects are closely involved with local communities, there is also often an agroforestry component.

Forest rehabilitation projects also extend to areas that have been destroyed by fire. There have been several projects in Indonesia, particularly on the causes and effects of fire damage, and it is largely as a result of this work that ITTO developed its *Guidelines on Fire Management in Tropical Forests*. Further work has been done since, inspired by the hugely damaging forest fires in 1997. Decision 8(XXIV) called for a technical consultation on forest fires in tropical forests and this took place in Indonesia at the end of 1998, following a study tour to Sarawak and Indonesia to assess measures of fire prevention. Fire is the subject of one of the case studies (case study 1 below).

The project 'Forest health monitoring to monitor the sustainability of Indonesian tropical rain forests' allows the integration of the use of criteria and indicators into national government schemes for forest monitoring as a means of assessing forest health. Until recently, such schemes have been tied to the concept of sustained yield rather than sustainable forest management. A few projects also address the problem of disease. Work has been carried out in Ghana, for example, on the development of pest management and genetic resistance in iroko (*Chlorophora excelsa*).

Sustainable forest management

Up to the end of 1999, there had been 71 projects in this subject area with the highest total investment of almost US$42 million. They include work on inventories, the design and implementation of management plans, the introduction of harvesting regimes, and the establishment of demonstration forests.

When ITTO adopted its first criteria and indicators, the Council invited consumers to make available funds for 'the establishment of demonstration areas for testing and demonstration of sustainable tropical forest management'.[3] As a result, there has been extensive activity, including projects in Tapajos in Brazil, von Humboldt in Peru, Chimanes in Bolivia, So'o lala in Cameroon, Hainan Island in China, and demonstration forest areas in both Sarawak and Peninsular Malaysia, Bulungan in Indonesia and Iwokrama in Guyana.

According to Secretariat estimates,[4] just under 1.5 million hectares of production forest were under ITTO-supported forest management at the end of 1999 – a figure which excluded conservation areas. This is a significant achievement, although the projects are at different stages and only a few are being harvested. Such areas could be used as a baseline from which to measure future progress.

ITTO's own initiative to set up a network of demonstration areas for the sustainable management of production forests has been less successful. Not all producer members were able to identify suitable areas and those that did lacked the resources with which to put the theory into practice. This has now been largely superseded by country-led initiatives.

Some assistance has been provided for training in RIL. By 1999, this was slight and not accorded the same importance in practice as in policy, but this has since been remedied by providing support for logging schools. ITTO has funded a RIL training programme in Brazil implemented by the Tropical Forest Foundation, the establishment of a permanent RIL training school in Guyana, and a RIL training programme in Cambodia. Projects to establish schools in other countries and regions are being developed; however, the Executive Director's proposal for regional-level training schools has not yet been taken up by members. ITTO does not currently have the resources to fund training schools in every producer country, so a regional approach would seem sensible.

Conservation

Some very significant projects fall into this group. The Mission to Sarawak in 1990 recommended that more land should be set aside as TPA; this has been followed up with the establishment of the Lanjak-Entimau Wildlife Sanctuary, through an ITTO project of three phases. The success of this project to preserve a substantial area of forest entirely for conservation purposes has been consolidated by a complementary project across the border in West Kalimantan. Together, the two reserves constitute a transboundary area covering over 1 million hectares. This is not only of great importance for the conservation of biological diversity, but also for monitoring and/or preventing illegal logging across the border. The concept of this successful joint initiative has now been replicated in other ITTO projects (see case study 2 below).

An early report on 'Realistic strategies for the conservation of biological diversity in tropical moist forests' initiated ITTO's work in this area and formed the basis of ITTO's subsequent policy publication, *ITTO Guidelines on the Conservation of Biological Diversity in Tropical Production Forests*. Despite this, only one project so far has really implemented the use of these guidelines at field level outside National Parks. Through the project 'Conservation and maintenance of biological diversity in tropical forests managed primarily for timber production, Surigao del Sur, Philippines', attempts are being made to steer practice in government-monitored plots away from an emphasis on sustained yield; the project has guidelines to revise logging management plans in order to incorporate the conservation and monitoring of existing biodiversity within concessions.

Community participation

These projects usually involve attempts by implementing agencies to work in collaboration with local communities to turn around a situation where forest areas have been encroached upon and degraded. Peru, Panama and Cameroon have several such projects where communities are working to rehabilitate the forest for their own long-term benefit. For many of these projects nursery establishment and agroforestry activities are included, as is the cultivation and use of non-wood forest products (see case study 3 below).

The importance of community and stakeholder involvement in forest management has become increasingly recognized during the life of ITTO; many developed countries give a high priority to the inclusion of social components in their ODA-funded forestry projects. Consequently, this type of ITTO project has been relatively successful in winning support and funding from consumer countries.

Distribution of funding

The distribution of the funds that have been available to ITTO has been wide in terms of benefiting 33 different developing country members, and also in terms of the range of projects, as illustrated by the different subject areas. However, funding of actions through the Special Account has lacked a coherent strategy because it is, necessarily, a political process. The financing of projects and activities that receive the Council's approval is entirely dependent on whether representatives of the donor community decide to contribute to them. And the donors are, inevitably, influenced by their own national aid policy priorities which discriminate against or towards specific countries and areas of activity. With an overall fall in ODA contributions from the developed world over the past decade and a drop in the number of countries regularly pledging funds, ITTO is increasingly constricted in what it can do and achieve; the net amount of new funds made available has remained relatively constant over the last decade when increases were needed. At each Session of the Council, the decision related to projects urges consumer members to commit at least 10 per cent of their contributions as unearmarked funds to the Special Account but this has never happened.

The result has been the piecemeal and inconsistent funding of projects. Feasibility studies have been implemented but the resulting project proposal has not. In some cases it seems that pre-projects have been funded simply because they have a smaller budget. Phased projects have been discontinued before they are complete. For example, Gabon started a demonstration forest project, the first phase of which involved the design of a management plan; the second phase was never funded and the plan has not been implemented. In contrast, the three-phase 'Demonstration program of sustainable utilization of tropical forests by means of differentiated management in Hainan Island' in China (US$3.3 million) is being implemented in spite of the fact that China is a consumer member and a net importer of tropical timber. Moreover, there is now a logging ban in Hainan, which prevents the application of the 'demonstrated' management regime. The commitment of the donor community needs to be clear – if a pre-project is funded, it should be assumed that the donors are, at least, interested in the work and likely later to contribute funding for the full project.

Conclusion

ITTO projects have certainly had a considerable influence in promoting the practice of sustainable forest management throughout the ITTO producer countries, but ITTO could have done much more and done it more efficiently. The Working Paper for the Yokohama Action Plan makes some interesting suggestions. These are along the following lines.

A national profile should be produced (much more simple than the criteria and indicators but encompassing trade and industry); requirements for funding must depend upon country characteristics such as those in Table 16.1; an objective system is needed to judge the relative importance of proposals to ITTO; there should be a better balance between divisions and a clear decision about the scope to be covered.

The last point is contentious. The Expert Panel's recommendations would have excluded many of the non-timber issues (conservation reserves, rural development activities, mangrove forests and non-timber products) from the Action Plan, but the Council wished them to be reinstated.

Clearly, the potential scope and influence of the ITTA is still open for serious debate. Some members want to interpret the mandate of ITTO narrowly, to limit its role and influence and to use other processes such as the UNFF and the CBD to promote non-timber aspects of sustainable forest management; others disagree. It would seem that the present climate of opinion among members is in favour of a wide interpretation of ITTO's remit, for ITTO has been able to develop a broad project portfolio in 'the conservation, management and sustainable development' of tropical forests.

Case Studies

Case study 1 – ITTO and fire management

ITTO became involved with forest fires soon after it came into being, examining the problems of rehabilitation after the very extensive and serious fires which ravaged East Kalimantan in 1986 and 1987. But it was the realization that fire was a major issue for tropical forests and must be dealt with in an integrated way that led ITTO to begin work, in 1993, on developing guidelines for managing tropical forest fires. Eventually published in 1997 as the *ITTO Guidelines on Fire Management in Tropical Forests*, these guidelines provide a step-by-step process by which tropical countries can analyse their fire management situation and develop workable programmes to deal with it. Written to be broadly applicable throughout the tropics, they contain seven broad categories of issues important to any fire programme: policy and legislation; strategies; monitoring and research; the institutional framework and capacity development; socio-economic considerations; land resources management and utilization; and training and public education. Within each category, 'principles' are listed that are known to affect fire management, and for each principle there are recommended actions. Those using the guidelines must evaluate the local situation and decide whether the recommended action should be applied as it is, modified to fit, or rejected as non-applicable. The value of the guidelines was recognized by the World Health Organization (WHO), which developed its *Health Guidelines on Episodic Vegetation Fire Events* with reference to the ITTO guidelines.

Indonesia

ITTO is working with several individual governments to implement the fire guidelines and to build capacity for managing fire. Indonesia, for example, was the first ITTO producer member to develop national guidelines on the protection of forest against fire, which it did with ITTO assistance and based on the ITTO guidelines.

The next step was to improve in-country fire management capacity. ITTO financed a series of training workshops for field workers across Indonesia. They included formal presentations on subjects such as fire prevention, fire behaviour and fire suppression, followed by practical exercises in the forest. Teams assessed the fire threat to particular areas, designed strategies to deal with what they observed and did hands-on work with the construction of fire lines and systems for the delivery of water.

Nevertheless, Indonesia recognizes that its capacity to manage its forest fires is still lower than is desirable and that much more work is needed. The Government of Indonesia and ITTO are developing a project to create a fire management service in the Province of West Kalimantan. There is high-level support for fire management in the Province and a commitment to involve all those likely to be affected – from concessionaires to local farmers.

Ghana

Ghana lost 1.3 per cent of its forest land (almost 590,000ha) between 1990 and 1995; the most important cause was forest fire. The Ghanaian Ministry of Lands and Forestry sought assistance from ITTO to bring the situation under control.

Using the techniques described in the *ITTO Guidelines on Fire Management in Tropical Forests*, the Ministry designed a fire management programme dealing with its fire situation, and ITTO provided the funds to begin implementing it.

If fire management is to be effective, it is imperative that local people understand the severity of the situation, and sympathize with and participate in efforts to correct it. The ITTO project has, accordingly, undertaken a broad programme of education and awareness in the prevention of wildfires. Jim Dunlop, a former head of the Forest Protection Branch of the British Columbian Forest Service, recently evaluated progress in Ghana's efforts to improve its fire management and assessed them as follows:

> Community work undertaken by the Ghana National Fire Service, the Forest Research Institute of Ghana and the Forestry Commission, and fostered by ITTO, has obviously convinced at least some of the local communities of the value of fire prevention. Further initiatives will likely find fertile ground for prevention efforts and it appears a very important change has taken place since the early 90s when no one seemed to understand or support the value of fire prevention at the community level.

An institutionalized fire reporting system is also being developed under the ITTO project. According to Dunlop, 'of all the functions or phases of a modern fire management system, none is more important than a timely and comprehensive database. Ghana is well on its way to building such a system.'

The future

It is difficult to create an integrated forest fire management programme. Governments can find the prospect so intimidating that they put off even making a start. When this is the case, ITTO has shown that it is able to work with national or sub-national agencies for natural resource management to develop such programmes, by establishing wide partnerships involving other international agencies, non-governmental and academic institutions and local communities. Drawing on its past experience in this area and its knowledge of global expertise, ITTO can build surprisingly effective projects from relatively modest funds. The Secretariat is now working with several of its members to develop a broader suite of projects in integrated fire management.

Case study 2 – ITTO and totally protected areas

In the early 1990s, ITTO worked with the IUCN to develop the *ITTO Guidelines for the Conservation of Biological Diversity in Tropical Production Forests*. These set out the reasons for making the conservation of biological diversity a goal of national forest policy and they show how to establish a PFE that integrates conservation areas with natural and planted production forests.

The guidelines provide advice on planning at the landscape level, such as linking reserves with corridors of natural forest to allow wildlife to move between reserves. At the field level they present principles and actions to maximize the conservation of biodiversity in the course of forest management.

The need for totally protected areas

ITTO is working with its partners on many field activities throughout the tropics to implement these guidelines. It recognizes that a network of TPAs – land dedicated primarily to conservation – is needed in addition to well-managed production forests. The ITTO biodiversity guidelines stress that the contribution of production forests to biodiversity conservation can only be fully realized within an integrated national land use strategy that includes such a network.

Accordingly, the Organization is providing funds for a growing suite of projects aimed at creating and maintaining TPAs (see Table 16.3). ITTO brings neighbouring countries and donors together at regular intervals. It has made use of this to establish a substantial programme of 'transboundary' reserves that straddle the borders of two or more countries. A main purpose of these transboundary conservation projects is, of course, to conserve wildlife but they also seek to improve the livelihoods of forest communities, promote cooperation between neighbouring countries, and control illegal logging and wildlife smuggling.

The first of the reserves financed by ITTO was the Lanjak-Entimau/Betung Kerihun Transboundary Conservation Reserve on the island of Borneo begun in

Table 16.3 *ITTO-funded transboundary and other conservation reserve initiatives,*
March 2002

Initiative	Partners	Funding countries	Area of influence*
Condor Range (Peru and Ecuador)	Ecuador Ministry for Tourism and the Environment; NATURA Foundation; Peruvian National Institute for Natural Resources (INRENA); Conservation International; local organizations	Japan, Switzerland, USA, Korea	2.42 million ha
Tambopata-Madidi (Peru and Bolivia)	INRENA; Bolivian National Service for Protected Areas (SERNAP); Conservation International; local organizations	Japan, Switzerland, USA	2.85 million ha
Phatam Protected Forests Complex (Thailand)	Thai Royal Forest Department	Japan, Switzerland, USA, France	130,000ha
Buffer zone of Kaeng Krachan National Park (Thailand)	Thai Royal Forest Department	Japan, Netherlands	348,000ha
Lanjak-Entimau/ Betung Kerihun Transboundary Conservation Reserve (Malaysia and Indonesia)	The Sarawak Forest Department; Park Management Unit of the Betung Kerihun National Park; WWF (Indonesia)	Japan, Switzerland	980,000ha
Kayan Mentarang National Park (Indonesia)	Directorate General of Forest Protection and Nature Conservation; WWF (Indonesia)	Switzerland, Japan, USA	1.4 million ha
Buffer zone of the Nouabalé-Ndoki National Park (Congo)	Wildlife Conservation Society; Government of Congo	Japan, Switzerland, USA, France	1.69 million ha (national park + buffer zone)
Mengamé Gorilla Sanctuary (Cameroon)	Directorate of Fauna and Protected Areas, Cameroon Ministry of Environment and Forestry	Switzerland, Japan, USA	137,000ha
Cahuinarí National Park (Colombia)	Colombian National Institute for Renewable Natural Resources and the Environment; Puerto Rastrojo Foundation; Bora-Miraña Indigenous Peoples	Austria, USA, Denmark, Norway	600,000ha
Iwokrama Forest (Guyana)	Iwokrama International Centre for Rain Forest Conservation and Development; Indigenous communities	Japan, Switzerland, USA, Korea	371,000ha
Total			10.9 million ha

Note: * Area of influence includes, in some cases, buffer-zone management areas.
Source: ITTO. Prepared for this book.

1994. ITTO is also collaborating with WWF (Indonesia) in providing funds for the management of the Kayan Mentarang National Park further to the northeast.

In South America, there are ITTO projects to establish a 2.9 million hectare reserve in the Tambopata-Madidi region on the border between Peru and Bolivia, and a 2.4 million hectare reserve in the Condor Mountain Range on the border between Peru and Ecuador. Both are funded by ITTO and are being implemented by the NGO Conservation International in collaboration with governments and local stakeholders.

In Africa, an ITTO project has led to the demarcation of a 500,000ha wildlife reserve in the Minkebe Forest in Gabon and has supported a pilot programme for the sustainable management of about 80,000ha of buffer zone. Another ITTO transboundary initiative launched recently is establishing a wildlife sanctuary on the border between Cameroon and Gabon in an area that is particularly rich in elephants, lowland gorillas and at least eight other primate species, as well as many other lesser-known species of plants and animals.

Buffer zones

Buffer zones round TPAs have long been recognized as an important way of reducing the damage that may be caused by human activities on TPAs. They were an integral part of the Biosphere Reserve concept developed in the United Nations Educational, Scientific and Cultural Organization (UNESCO) Man and the Biosphere (MAB) programme in the early 1970s[5] and IUCN produced guidelines for their management in 1991.[6] ITTO is working in this field also. In addition to the Minkebe Forest in Gabon, mentioned above, there is an ITTO project, implemented by the Wildlife Conservation Society and the Government of Congo, to protect the immensely important Nouabalé-Ndoki National Park in northern Congo. This park covers 390,000ha and is rich in wildlife, including the western lowland gorillas, chimpanzees and forest elephants. The project manages nearly 1.3 million hectares of forest adjacent to the park, using an approach that balances biodiversity conservation with timber production and other income-generating activities. The key to success in this project – and in most tropical forest management endeavours – is to engage the local communities and to help them to improve their economic, social and political status. For example, the project is employing forest guards recruited from local villages to control wildlife poaching; such employment gives an additional incentive to conserve wildlife. The logging company working in the area – also a partner in the ITTO project – is committed to minimizing logging damage and, assisted by ITTO, is training its workforce in RIL.

Another ITTO project is establishing a working model for effective buffer zone management at Kaeng Krachan National Park in Thailand's Petchburi Province, by working with the local community both to find ways of increasing the income of villagers without compromising the integrity of the park and to make them aware of the benefits from conservation. In Guyana, a 'sustainable use' zone has been established in Iwokrama Forest next to a newly created 187,000ha TPA. In it, the local indigenous communities are developing and implementing a sustainable management strategy funded by ITTO.

Case study 3 – ITTO and community forestry

Communities living in and around forests have traditionally used forest resources, either by harvesting wood and NWFPs or by clearing patches of forest for agriculture or pasture. If these forests are to be managed sustainably, local communities must take part in planning and implementation and must have a share of the benefits. Such community forestry – management with or by communities – is an important way to promote social equity while managing the forest sustainably. Some ITTO producer members have given high priority to community forestry in their national land use strategies and, in many cases, have obtained ITTO support through project funding. In fact, community involvement is a strong focus of at least half of ITTO's present projects in reforestation and forest management.

Community forestry may often be the best means to achieve sustainable management of forests but it is not easy to bring about. Perhaps the single biggest stumbling block is that the community has no security of tenure. For, without established and long-term rights, few people living on forest lands see their investment of time, labour and other resources to be of any lasting value. The most successful ITTO projects, therefore, are those that build on government initiatives to offer tenurial rights to communities.

For example, an ITTO-financed project in Nueva Viscaya in the Philippines supports a community-based forest management agreement between the Government and the local community federation, comprising three farmers' associations – the Kakilingan Upland Farmers Association, the Kalongkong Upland Farmers, Inc and the Vista Hills Upland Farmers Association. This agreement confers security of tenure and provides incentives to develop, use and manage allocated portions of forest lands pursuant to an agreed framework spanning a period of 25 years, renewable for another 25 years. It includes a production-sharing agreement designed to ensure that the benefits of sustainable use, management and conservation of forest lands and natural resources are shared equitably.

Another key to success in community forestry is continuity of support. Community forestry requires lengthy processes of participation and conflict resolution, which themselves must be supported by detailed background information, the education of those concerned and the winning of trust. ITTO projects are generally short – most commonly three years; but they can be extended. The Nueva Viscaya project began in 1993 and is now in its second phase. Long-term support is essential if community forestry projects are to be successful. For this reason it is still too early to assess the effectiveness of many ITTO projects in this field.

Community forestry is often hindered by a lack of industrial capacity; this requires some cooperation between forest-harvesting companies and local communities. Such cooperation is often difficult to foster and maintain, but ITTO projects can help. In Colombia, for example, land property rights are being granted to the rural black communities in the Lower and Middle Atrato Regions in the Department of Choco, in accordance with the provisions of Colombian Law 70/93. This law also encourages partnerships between the private sector and the local communities to use the forest resources of the region – once the communities have formally organized themselves into community councils. So far, eight community councils have been granted land rights in the Truandó-Domingodó region over an area of approximately 80,000ha. An ITTO project supports the first three of these associations – Acamuri, Asocomunal and Ocaba. It aims to establish and manage the first 2000ha of protection-production plantations in the region.

Within this project, the communities have chosen to work in close cooperation with Maderas del Darién SA, a local company with extensive experience in the proposed forest practices. The project also ensures the participation of the Choco Sustainable Development Corporation (CODECHOCO) – a regional government agency responsible for regulating the use of natural resources at the local level – that has agreed to act as an arbitrator between community councils and industry in order to guarantee fair play and an equitable distribution of benefits. The early success of reforestation carried out under this project has created enormous interest among other local communities, several of whom have requested the timber company, the Corporation, the municipal authorities and the Colombian Environmental Authority to implement similar programmes elsewhere.

There are also problems of intra-community conflict. In Bulungan, East Kalimantan, Indonesia, CIFOR is implementing an ITTO project to find better ways of looking at logging practices and integrating the social, silvicultural and economic aspects of long-term forest management. Village-to-village coordination is being examined – a subject that has received little attention but is fundamental – in particular how boundaries are demarcated between the villages of 27 Dayak (primarily Merap, Punan and Kenyah) villages in the catchment of the upper Malinau River. It found that most inter-village conflict centred around claims to agricultural lands (swidden fields, wet rice fields and perennial gardens) that, according to customary rules, rightfully 'belonged' to the household establishing the plot, even if they fell within the territory of another village.

The project facilitated participatory mapping in which villages negotiated boundaries with their neighbours. A team of villagers, with project assistance, next identified and mapped village boundaries; 21 villages completed negotiations and the mapping of their territories. The process unearthed considerable inter-village tensions that are difficult to resolve – but must be, if sustainable land use patterns are to be established.

Community forestry is a popular movement, heralded by some as an agent for democratization; imposing from outside is likely to be counter-productive. ITTO supports approaches developed within the target countries themselves, often implemented by small NGOs. For example, the Evergreen Club, an environmental NGO in Ghana, promotes the participation of primary-school children and their teachers in local tree-planting and reforestation; the increased awareness among the children of the need to protect and replant trees is transferred to their parents and other villagers. Another project, also in Ghana, run by the 31st December Women's Movement, assists women in several villages to establish and run tree nurseries, and to practise agroforestry in the degraded forest areas surrounding their villages.

Case study 4 – ITTO's role in promoting market transparency

A major stumbling block in developing and sustaining valuable and equitable markets for tropical timber products is the inadequacy of information about the trade. It is only through the free flow of reliable information on forests and trade that governments and industry can evaluate their resources, set goals for development and act to increase the trade, capture more of its value and discourage illegal activities. Information promotes transparency; and this is important for other reasons. Consumers increasingly seek evidence that the timber they buy has been harvested from well-managed forests. Equally, forest owners and people living in or near the forest want assurance that a fair share of the benefits of the timber trade are used to improve local livelihoods and to meet the costs of sustainable forest management.

In support of these needs, ITTO has from its inception devoted considerable resources to improving market transparency. Member countries have been assisted through projects to strengthen their statistical and data collection and reporting functions. In addition, the ITTO Secretariat regularly reports on tropical timber production, trade and market trends and investigates a host of market-related issues. Statistics on tropical timber production, trade and prices, maintained in computer databases at ITTO headquarters, are analysed and published each year in the *Annual Review and Assessment of the World Timber Situation*. This review contains comprehensive timber production and trade statistics at the global level, combining data collected from ITTO members (in cooperation with FAO and the UN Economic Commission for Europe (ECE)) with that obtainable from other sources such as the UN COMTRADE trade database. Data from ITTO's *Annual Review* are widely used by member governments and by both trade and environmental NGOs. Users of ITTO market data include such diverse organizations as the Organisation for Economic Co-operation and Development (OECD), the WRI, WWF, the Common Fund for Commodities, and Global Forest Watch. In their efforts to understand and address illegal logging and illegal trade practices, many such institutions (both in collaboration with ITTO and on their own) are analysing data from the *Annual Review* to identify anomalies in trade statistics as a possible indicator of failures in law enforcement and export control.

The *Annual Review* depends largely on statistical information supplied by members, but many countries, both within and beyond the tropics, need more help to deliver such information to an adequate standard. A system of nominating statistical correspondents in member countries was particularly successful and led to improved statistical information in several countries. A parallel development was the implementation of a series of ITTO projects under which 15 training workshops have been held in each of the three producing regions of Africa, Latin America and Asia. These workshops gave participants, in many cases their country's nominated statistical correspondent, hands-on experience in the collection, analysis and dissemination of forestry statistics, including sessions on completing the annual statistical enquiry received from ITTO. To date, regional workshops have been held in Malaysia, Bolivia,

Côte d'Ivoire, India, Brazil, Cameroon, Gabon, Colombia, Venezuela and Togo for about 450 trainees from over 30 countries.

To further strengthen the collection, analysis and delivery of tropical timber trade data, ITTO has responded to requests from member countries by funding specific national projects on trade transparency and the improvement of data collection and analysis. So far, the Organization has funded 26 such projects in 15 countries, focusing on those countries with the greatest need for improving their statistical responsibilities to the Organization. In Gabon, ITTO projects developed capacity for the collection and computer processing of forest statistics. In Peru, a project established a strategic forest information centre to collect and disseminate information on, among other things, the harvest and transport of all timber in the country. In Fiji an ITTO project developed an export market intelligence monitoring system to provide a chain of custody for timber exports. Other statistical development projects are now being implemented in Ecuador, Bolivia, Peru, Egypt, Gabon, Cameroon, China and Togo.

In addition to the annual statistical review produced by the ITTO Secretariat, the Organization's Tropical Timber Market Information Service (MIS) delivers bi-weekly market news, prices and trends for hundreds of tropical timber products to trade groups, executives and analysts world-wide. This is part of the effort by ITTO to promote a competitive timber industry and a transparent trade in tropical timber products. The MIS report publishes a range of data to meet producers' needs for information on trends and developments in the consumer markets as well as information from producer countries of interest to timber buyers.

ITTO has funded the development of models to simulate different trade scenarios given future changes in economies, usage patterns and supply. For example, timber-supply and trade models have been developed to help evaluate the medium and long-term outlook for the tropical timber market. A separate study recently forecast China's production, consumption and trade of forest products to the year 2010. This produced new data showing the rapid growth of China as a market for tropical timber with the expansion of its housing and furniture markets. It predicted that these markets would start to change significantly by 2010 as China's own plantations replace imports and as substitution of tropical timbers increases.

Market studies have been conducted in the major consumer markets, including Japan, North America, Korea, China and the EU. Other global studies have investigated the growing significance of trade in secondary processed wood products and, more specifically, the international wooden furniture markets.

The tropical timber trade and forest industry still face considerable hurdles if they are to maximize their contribution to sustainable development and tropical forest conservation; there is still a need for more information on trade, greater development of the processing sector, and greater access to consumer markets.

17

ITTO: agent of change

'Some criticize the Agreement because it has no teeth; there are no sanctions for non-compliance. But it is hard to see what form sanctions could take except, perhaps, in the case of illegal trade. It may be better to be toothless than gnashing teeth to no effect'

In this chapter I examine how useful ITTO has been in working towards the best conservation and use of tropical forests in the long-term interests of humanity. What have been its strengths and weaknesses? What, if any, have been its comparative advantages?

It is, of course, impossible in such a complicated field with so many players to identify which or who has had the greatest influence at any time. Often the first articulation of an idea is so far in advance of its time that nothing happens. Progress is usually brought about by a fortunate, and often fortuitous, coincidence of the right idea at the right time – and sometimes also the right individual. But one can, at least, look at the positive steps that ITTO has taken (Chapters 14 and 16) and reflect on the advantages and disadvantages possessed by ITTO as an agent for change.

Let us look first at the strengths. ITTO is an organization set up by treaty. The obligations of members have, therefore, a more formal and, one hopes, more binding basis than members would have in a non-treaty organization. The formal decisions reached should have special force – which is no doubt why so much time is spent in debating their exact wording. Another strength is the voting system; the way that votes are distributed gives equal rights to the producers and consumers of tropical timber – although, as we have seen, some 'producers' are in fact net importers of timber. In fact, the distribution of votes reflects, with a few exceptions, the distinction between those who have tropical forests and those who do not, and more-or-less segregates the developing and developed nations. So far ITTO has been able to reach all decisions involving policy or projects by consensus;[1] consequently the producers, who have been in a sense in the firing line, have always felt that they have ownership of the decisions and are more likely to honour them.

Two strengths of ITTO are that it links the forest resource, its management and trade in its products; and that it can treat forest management in the widest context of land use and social and economic relationships. It is, of course, not unique in this: the terms of reference of FAO cover all these topics and FAO has consistently been recognized by the UN as the lead agency in forestry matters. Hidden in the record somewhere one might be able to disentangle why it was that governments felt it necessary to set up a new organization rather than remit this special task to FAO.

From the very beginning, ITTO has opened its ranks to NGOs of many persuasions – from various branches and complexions of the environmental movement, from the trade and from scientific and other bodies; it is unusual for an intergovernmental organization to be so open. We shall comment below on the very different uses that these groups have made of this privileged access.

One influential feature of ITTO is that it is at the same time a forum in which forest policies may be discussed – and the discussion has often been lively and sometimes contentious – and norms and procedures may be established, and an agency that can approve and finance projects. This again is not unique, though few organizations would have been able to establish something with such powerful appeal as the Year 2000 Objective. But what is perhaps unique is the flexibility of ITTO in setting priorities (for missions, field operations, training, fellowships, etc – less so for projects that require earmarked funds from donors) and the speed with which it can translate decisions on policy into action on the ground. Where there have been weaknesses in this sequence – for example in the quality of some projects, delays in implementation, lack of funding and uncertain focus – these are remediable faults in procedure rather than inherent in ITTO itself. And much effort has gone into trying to correct these deficiencies. There has always been some possibility for the Secretariat to stimulate studies, activities and projects in particular fields but this has now been much strengthened by the sending of diagnostic missions to countries, on request, to identify the factors that most severely limit their progress.

One early weakness of the ITTA, 1983 has been removed in its successor. While the resource was entirely located in producer countries and its management was ultimately their responsibility, the trade was subject to much wider influences, notably competition from non-tropical timber and industrial substitutes for wood. ITTO now has the very great advantage that it can discuss all the global trade issues that affect the profitability of forest management and the timber industry while concentrating its projects on the tropical forests – where the need is greatest. Its focus can be on the issues of greatest moment – the sustainable *and* profitable management of tropical forests, equitable trade and an equitable distribution of benefits.

Finally, ITTO has a small secretariat – probably slightly too small to cope with such a wide range of duties – but one that can operate with a minimum of red tape.

Nevertheless, there are a number of countervailing disadvantages and weaknesses. Some criticize the Agreement because it has no teeth; there are no sanctions for non-compliance. But it is hard to see what form sanctions could take except, perhaps, in the case of illegal trade. It may be better to be toothless than gnashing teeth to no effect. It is evident that, during the life of the Agreement, it has helped to establish norms that are now broadly accepted, and it is slowly coming to be a question of national prestige to conform to these norms. The projects act as an additional incentive.

Another area of uncertainty is whether it is a benefit or a hindrance to divide the Council formally into the two caucuses of producers and consumers, especially now that the distinction between the two is becoming blurred. This arrangement does allow the members of each group to iron out their differences in private and establish a common position but, on some occasions, it has undoubtedly led to unnecessary confrontations.

The Council has also shown itself to be frustratingly ineffectual in certain areas. A good example is the decision to review progress towards the Year 2000 Objective based only on country reports: if the Year 2000 Objective has been the Organization's most important policy initiative, its inability to properly assess progress towards it must count as its greatest failure.

The regular market discussions at each Council session have provided an admirable forum for a frank exchange of views but, after the first few sessions, there has been no comparable forum to discuss questions of reforestation and forest management. This is a pity. Instead, there has been the clumsy and expensive expedient of expert panels where the choice of an equal number of members from producers and consumers has tended to produce groups of negotiators rather than of experts. Finally, at the level of the Council, such issues tend to be settled in two stages – first by the formation of positions in the caucuses and then by using the drafting sessions for the actual formal decisions. The result has been time-consuming, inefficient and acrimonious.

These, however, are all difficulties of procedure rather than weaknesses inherent in the Agreement. The same can be said of deficiencies in the formation and choice of projects and the unwillingness of Council to delegate responsibility to the Executive Director and the Secretariat to the degree that is normal in most organizations. These weaknesses are recognized and steps are being taken to correct them.

More fundamental is the shortage of funds to make sustainable management a reality in producing countries. There is argument and contention about the partition of benefits between those that are global and those that are national. There is no doubt that *all* measures for conservation benefit *everyone* – every country is part of the global community. Many of the advantages of sustainable forest management *do* accrue to the producer countries themselves. Among these are stability in timber production,[2] more reliable provision of good quality water, freedom of rivers from siltation and pollution, and some elements of biodiversity. But there is equally no doubt that some benefits are more global than others, notably carbon storage[3] and the general values of preserving biodiversity. Where the costs fall mainly on the poorer nations, the richer producer countries should be prepared, and some of them are prepared, to make a disproportionate contribution towards the conservation of these values.

It is abundantly clear, however, that the poorer countries cannot afford to make the necessary investment in organizational capacity, trained personnel and infrastructure to bring their forests under truly sustainable management without greatly increased funding. *Its inability to attract funds on the scale required is the fundamental weakness of ITTO.*

As has been noted above, one of the potential strengths of ITTO has been its openness to the opinions and influence of NGOs, of which the main groups have been either environmental or associated with timber trade and industry. These have operated in rather different ways. The trade representatives have on the whole been cohesive and, by setting up regular market discussions, have been able to keep topical issues such as the downturn in the market, tariff barriers, certification, etc under review. Their role, too, has been in the shared economic interest of producers, consumers and the trade, and close to the original reason for the Agreement in the first place. They have generally been interested in operating within the confines of the Agreement.

The environmental NGOs have been a more disparate group. They have included international pressure groups such as FoE and Survival International interested to different degrees in sustainability, biodiversity and the rights of indigenous peoples; complex associations such as IUCN with many members both governmental and non-governmental; WWF, containing both a central international organization and many country-based branches; IIED with its interest in *both* environment and development; and, especially at the meetings held outside Yokohama, many local NGOs – often those associated with ITTO field projects. It is hardly surprising that this heterogeneous group has shown less cohesion than the trade. In the early meetings, all had high hopes and worked together in providing a mixture of helpful advice and criticism – for example, proposals for the better orientation of projects (WWF), the proposed project on labelling (FoE), and an environmental newsletter. IIED, with its survey of the extent of sustainable management (Chapter 5) and other initiatives, believed that it could have more influence if it continued to work within the parameters of the ITTA.

There was, thus, an early division into those who, to different degrees, believed that it was most productive to work within the formal limits of ITTO and those who wished to change the system. As we have seen, the latter were disappointed and withdrew. In Sarawak the clash of interest between WWF (Malaysia) and WWF (International) was along these lines. It is possible that the unfortunate divorce between ITTO and the international environmental NGOs[4] in the early 1990s might have been avoided if a forum had been set up, parallel to the market discussions, to engage in constructive debate about the issues raised by the idea of sustainable forest management. Very recently, environmental NGOs have been showing renewed interest in ITTO, perhaps because of its new dynamism and the willingness of the present Executive Director to engage in open dialogue. Perhaps, too, they have become disillusioned with the alternative international processes.

The history of the environmental NGOs within ITTO raises a very interesting and perhaps unanswerable question. Is it better to reform from within by undertaking the achievable, or to break the system in order to make possible what was previously unachievable? Is consensus not more constructive than confrontation?[5] ITTO provides a fertile ground for working out these relationships and tensions.

Changing landscapes, future prospects[1]

'ITTO was first devised to address a specific concern but it has evolved into something much broader; in the process, negotiators have created something new, and the lessons to be drawn from it are many'

The world is changing rapidly. We can only guess what it will be like within the life of trees that are now seedlings. Shifts in the number and composition of populations; migration to cities; greater mobility, communication and trade; higher expectations and consumption; changes in climate; new technologies and sources of energy will create a global landscape that may well be unrecognizable to the present generation. There is no certainty of peace and security. More than ever before, it is necessary to remain flexible and to plan to adapt to inevitable change.

What has this to do with forests? In many parts of the world, natural forests form the climax vegetation. They have been shaped by their history, adapted over millennia to past changes in their environment. They are thus especially capable of responding to future change – provided, of course, that this is not too drastic or too rapid; the less they have been altered, the more adaptable they will be. Forests have other advantages too: they and their soils store carbon – the higher the standing volume, the greater the storage. Furthermore, wood is a renewable resource, carbon-neutral and environmentally friendly. Therefore, other things being equal, the more forest we have and the more we use wood rather than energy-greedy metal, plastic or concrete, the better equipped we shall be to face the future.

The sustainable management of the world's tropical forests depends on two broad trends, the future patterns of land use in tropical countries and the future management of those forests – in fact, what happens *to* the forest and what happens *in* the forest or, to phrase it differently, what forests we keep and how we manage them.

Much attention has been given in recent years to deforestation and its causes rather than to deciding what forests the world needs and how to ensure that they are preserved. This obsession with deforestation has not been helpful. It seems to be based on two premises: that the removal of forest is always a bad thing and that any increase in forest cover is good. These premises are not universally valid either from the viewpoint of social development or even on environmental grounds. In writing this, I am *not* arguing in favour of forest destruction; for example, there is no doubt that when some types of tropical forest are removed, the resulting ecosystems are very much less rich in species and less productive than the forests they replace. But I *am* arguing that we should stop shedding crocodile tears about the loss of forest *in general* and direct attention to the more important questions of how much forest we need or want, what kinds of forests these should be, where they should be situated and how they should be preserved and managed.

How Much Forest?

It is very much easier to ask this question than to answer it; neither is there any universally valid recipe for reaching an answer. It is for each country to assess for itself how much of its land area it wishes to be forested, taking into account the various values of forest, the other uses to which the forest land might be transformed, and the opportunity costs of retaining it as forest. Although the most pressing concerns in any country are local and national, any responsible government will also take note of global concerns.

At any time – shall we say 'the present'? – any nation or large self-governing unit has a certain area of forest (of different forest types, in various locations and used for different purposes). It might wish that it had more forest, that the forest was distributed in a different pattern or that it was used in different ways. It might even wish that there were less forest! But each nation or province has to start with what it actually has. I suggest, however, that it is very useful, indeed almost essential, for each nation first to review in broad outline the present status of its forests – their area, kind, quality, ownership and location – and then to assess the probable future demand for all purposes: for the production of wood and other materials, for environmental reasons, and for diversion to other purposes such as agriculture, infrastructure and urban development. Such a general survey of forest resources, and of the demand for them, should enable the nation to make an approximate assessment of how much forest it needs (and for what purposes and where). The very process of going through this exercise should ensure that all the possible values of forest are taken into consideration and duly balanced one against another.

Special care should be taken to avoid changes that are irreversible or very expensive to reverse – the extinction of species, the disruption of indigenous communities, the disappearance of the last examples of particular natural ecosystems or severe deterioration of soil and water. Moreover, until the time comes to exploit or convert, it is wise to retain forest in as natural a state as possible, thus retaining a wider and cheaper range of choices of future use (Box 18.1).

The selection and establishment of a permanent forest estate along these lines is perhaps the most important single measure that a country can take to preserve its forests for the future and to provide a buffer against future changes.

BOX 18.1 RELATIONSHIP BETWEEN THE LEVEL OF FOREST DEGRADATION, THE MAGNITUDE OF CHANGE, THE RECOVERY PROCESS AND THE COST OF CURATIVE ACTION

The sensitivity of forest ecosystems to stress is determined by the type of forest ecosystem (resilient or fragile) and by the form of stress (type, duration and intensity). A generalized response is illustrated below. Forests subjected to stress beyond their tolerance may follow the F1–F2 trajectory. The recovery pattern may be divided into three levels:

1 *Self-renewal.* If degradation is moderate and stress is withdrawn, forest ecosystems can renew themselves, returning relatively quickly to more or less their original state without human intervention.
2 *Rehabilitation.* At an intermediate level of degradation, the forest ecosystem may take a long time to recover naturally, but this may be shortened by human intervention.
3 *Restoration.* At a certain level, forest degradation becomes practically irreversible. It is characterized by a total or near total loss of forest cover and of characteristic forest species as well as by soil degradation and reduced productivity. Recovery may take centuries if it depends only on natural processes. It is impossible to recreate the 'original natural' state of the forest; but it may be possible to create a partial substitute.

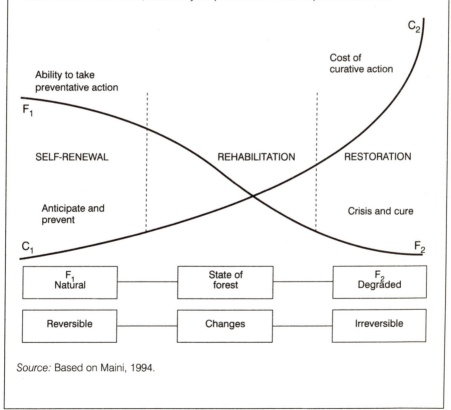

Source: Based on Maini, 1994.

The figure also suggests a number of other conclusions:

- Preventive action that costs little or nothing can only be taken in the early stages of forest degradation.
- The cost of cure increases with the extent of forest degradation (trajectory C1–C2).
- It is important to recognize early warning signals and to 'anticipate and prevent' rather than 'degrade and cure'.
- We are better able to practise sustainable forestry if we can predict how natural disturbances and human activities affect forest ecosystems.
- Carefully designed, long-term studies will make us better able to predict.
- It is important to set aside ecological reserves of representative and unique forest types as environmental baselines.

Trends in Forest Use

Polarization

Let us now speculate about the trends that may affect the existing areas of tropical forest, for some judgement about these trends will influence decisions about the choice of areas of forest to be retained or about the need for new forests.

Much of the expected growth in population in the next half century will take place in tropical countries. They will wish to transform some forest on suitable soils into farmland to grow food or cash crops, and some forest, too, will be required for new settlements and infrastructure. This is inevitable; it is better to plan for it than to resist. If the lands are well chosen and the conversions are carefully planned and conducted, less forest will be harmed.

Changes in the condition of rural peoples should be anticipated. Take the indigenous peoples of Sarawak, Malaysia, as an example, now increasingly affected by contacts with the market economy and the values of the metropolitan society. While the older generation, whether of hunter-gatherers or shifting cultivators, often wishes to continue the traditional way of life and needs the space and freedom to do so, the younger generation has mostly left their traditional longhouses to work in the larger towns, returning only for weekends and holidays. Swidden agriculture is carried on by their parents, often now employing paid immigrant labour. If they, and shifting cultivators like them in other countries, follow the sequence of events that has been almost universal in richer countries, shifting cultivation in the forest may cease in a few generations. Moreover, many other changes may be expected as rural peoples gain confidence in claiming rights to the land they occupy.

The present trend seems to be towards a polarization of land use towards intensive agriculture and plantation forestry on the one hand and protection and non-consumptive use on the other. Both are relatively simple concepts, easy to manage, and therefore attractive to planners and administrators. If compared with the sustainable use of natural ecosystems (whether as forest, pasture or for a crop of wild animals), intensive use occupies less land and brings greater financial returns, while protection is relatively cheap and has been shown to

Box 18.2 Relation between yield and the area of forest needed to meet targets for world timber production

There are very great differences between the annual production of wood in natural forest and in plantations, between the same natural forest under different intensities of management, and between plantations of different species. The curves below show the areas of forest that would be needed to produce 3000, 4000 and 10,000 million m³ of timber per year under different kinds of management and illustrate that the world's requirements for wood could in principle be met from a fraction of the present areas of forest.

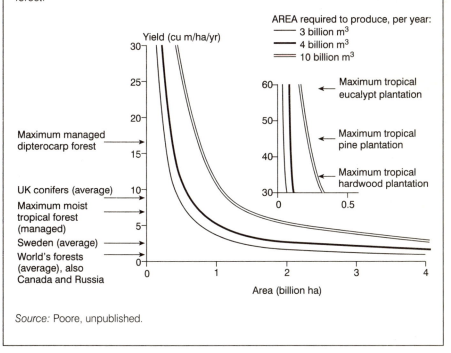

Source: Poore, unpublished.

work. Both, in an appropriate land use setting, have proved highly effective for their respective purposes (Boxes 18.2 and 18.3).

Intensification in agriculture and forestry is generally market-driven. The motivation for protected areas is more complex. Where it is a matter of protecting water catchments, the local benefit is obvious. The conservation of biological diversity is rather different. It is often portrayed as being mainly for the benefit of the global community, but this is only partly true. Any country is itself part of the global community and it is likely to gain at least as much benefit as others from the conservation of its own biological diversity. It is general experience that, as development proceeds, conservation policies are stimulated by the public opinion of an increasingly aware electorate. The conservation of biological diversity will also become a suitable use for remoter forest areas from which economic activity has withdrawn.

BOX 18.3 THE CONTRASTING EFFECTS OF DEFORESTATION AND PROTECTION OF BIOLOGICAL DIVERSITY

1. The species area curves often used in predicting extinction rates are based on the comparison of geographic entities (often islands) of different areas. They are therefore generalizations obtained by including the whole variety of communities in these entities. The rate of extinction for any loss of area is often assumed to follow these curves. This may give misleading results, which can usually be explained by examining in more detail the species/area curves of individual communities.

2. The following arguments are based on data from communities of higher plants, bryophytes and lichens; it is thought that they also apply to micro-organisms and sedentary animals.[2] Almost all plant communities examined show species/area curves rising rapidly to an asymptote (known in the literature as 'minimal area curves').[3] The communities differ from one another both in the total number of species occurring in them and the area at which the curve tends to flatten out. They also differ in the *total* area that they occupy. At the two extremes, shown in the diagram below, are local communities with many species (A) and extensive communities with few species (B); in between are many intermediates. Only in the case of certain tropical forest types has the situation been found in which the uniform areas available are not large enough to accommodate all the species that could potentially occur upon them (eg Jengka, Malaysia).[4] The combination in the same data set of a large number of these individual community curves produce the generalized curves mentioned in paragraph 1.

3. Consideration of these curves from individual communities can provide a large part of the theoretical basis of understanding the observed rates of species extinctions and for planning conservation measures.[5] For example, a large proportion of the area of community B can be destroyed without causing any extinctions, whereas this is not the case with community A. Conversely the preservation of relatively small areas of each community should safeguard the large majority of species belonging to it.

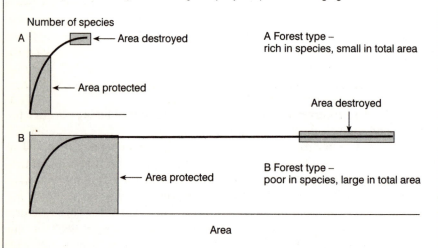

Species/area curves for species-rich and species-poor forest types

Source: Poore, unpublished

Intensive systems of production and the markets they serve are vulnerable if energy ceases to be abundant, convenient and cheap. The more intensive the system, the more dependent it is on the use of machines, fuel and fertilizers. Moreover, in the present-day market society, the site of consumption of many products is often far from the site of their production.[6] There is little doubt that intensive systems will be able to adapt to new circumstances but the process may be difficult and painful.

Sustainable use

The sustainable use of natural ecosystems such as natural forests and rangelands lies in the middle ground between intensive use and protection. The management of these to provide a sustained yield is inherently more complicated than that of annual field crops or of tree plantations. The primary aim of management is often to provide an economic yield, but the vegetation may also perform ecological services – as protective ground cover or a reservoir of biological diversity – in which case the objective may be to provide the best durable mix of goods and services. Tropical forests are particularly complicated because of the abundance of species, their relatively slow growth rates and the way in which demand for their products changes with time.

The usual view of the management of tropical forests for a sustained yield of timber is this (see also Chapter 2). A regular harvest from any compartment of the forest is taken every so many years, the length of rotation depending upon the growth rate of the harvestable trees. The forest is then closed until the harvest of the next crop, in perhaps 20 or 30 years. If the area of forest is large, the harvest can be averaged out so that the forest as a whole produces a regular annual return; if the area is small this is not possible.

In many different kinds of tropical forest, such management has been shown to be technically possible – and not as complicated as is often made out; but ITTO studies, among others, have demonstrated that it has rarely been carried out consistently even for one rotation. Why, then, has it proved so difficult? Almost certainly because it is rarely profitable to do so. The profit to the owner (or concessionaire or manager) depends largely on the market price of the goods he produces, unless he can also receive payment for the forest services provided through his good management. This is so whether those who stand to profit are governments, companies, individuals or local communities.[7] The first cut from previously unexploited forest – the mining of timber – is of course profitable, but only very occasionally has part of the profit from this first cut been reinvested in future management. Thereafter, the returns rarely justify the expense of careful management. Tropical forest is being cleared for other uses because the land brings the prospect of a greater financial return when used for other purposes than it does under management for the sustainable production of timber.

The sustainable management of natural tropical forests for timber is not only less profitable than other uses to which their land may be put; it is less profitable than tropical plantations or the sustainable management of temperate forests. Even if profits are increased by higher efficiency, increased yields, added value, better marketing and the possible advantages brought by certification, it seems unlikely that tropical timber will remain competitive in the market and bring sufficient return to entice investment. Higher costs of labour or energy may make it even more difficult; so may any more stringent environmental or labour conditions imposed by the market.

Direct payment for the provision of environmental services would help, but payment by whom? The developed world seems more ready to pay for conservation than for sustainable use. Does this mean that sustainable management for timber PFE must be abandoned in favour of protected areas?[8] Is there not still room for both? Is it not possible that the whole notion of managing natural forest to provide a *regular* return is itself wide of the mark? External conditions (such as labour, markets and environmental conditions) are changing so rapidly that it seems unrealistic to produce management plans that assume that there will be a sustained yield of the same products over a number of cutting cycles or rotations.

There is another possible model, which may be called 'adaptive management'. This is to give a grand name to something that has been practised by nomadic pastoralists, for example, for centuries – perhaps millennia. It is, in effect, to fit management operations to the needs of the moment. Private forests in Europe, where there is a centuries-old tradition of forest management, are seldom managed to produce a regular income. They are managed, instead, as accumulating capital assets that can be tapped when there is a need to raise capital (to put a new roof on a house or provide a dowry for a daughter) or when there is a special market opportunity. Much farm forestry in the tropics is managed on the same principles. I suggest that this is the possible way forward for the sustainable use of tropical forests. A prerequisite will be secure tenure, which might be by governments, companies, private individuals or local communities. In contrast to the initial phase of mining the forest, this system of management would draw from the forest what was required for local use and what was marketable at any time – including, for example, high-value timbers. It would retain the structure and diversity of the forest and thus its flexibility. Rather than forest capital being dissipated, it would be encouraged to accumulate; at the same time, the increase in standing biomass would make a contribution to carbon storage.

The Way Forward: from Words to Action

Wrong directions?

During the past 30 years there has been huge investment in forestry projects and many millions of words have been spoken and written on the subject. The ITTO Year 2000 Report and recent FAO and UNEP/WCMC assessments show that there has been some progress – in laws, policies and institutions; in the establishment of a PFE comprising forests for production, protection and conservation; in the involvement of communities – and in the production of management plans. There is a growth industry in the generation of norms – guidelines, criteria and indicators and standards – and in the development of systems for certification and auditing. But the area of tropical forests that are demonstrably sustainably managed for timber is still very small. Moreover, in spite of its undoubted value in creating mutual understanding, the IPF/IFF/UNFF process has produced few recommendations for action which have anything to do with action on the ground – either sustainable patterns of land use or sustainable systems of management. In fact, most of the action has been peripheral – displacement activity.

Although it has long been recognized that forestry is an isolated profession and that the problems of deforestation and the degradation of forests mainly lie in policies outside the forestry sector, forests continue to be treated as something special and apart. This dangerous segregation, which was reinforced at Rio, would become institutionalized and to all practical purposes irreversible if there were to be an international convention devoted exclusively to forests.

Because of the particular focus directed towards forests and especially tropical forests – largely by the environmental NGOs – far more stringent requirements are expected for forest products than are expected of products that compete with them, such as plastics, concrete or steel. More exacting conditions are placed on forestry than on agriculture. Those who call for them do not understand the management of forests; nor do they allow for the possibility of genuine and remediable mistakes.

The calls for bans and sanctions in the late 1980s and the immoderate criticism of the actions of tropical forest countries did much to polarize the debate at Rio and made forestry a much more contentious subject than was justified by the facts. It set back a reasonable and constructive reconciliation of differences for at least a decade.

The ideas of certification and of forest standards are excellent in intention; they were inherent in the first proposal for labelling that came before ITTO in 1989 but, if the way they are being developed and implemented inhibits and stultifies sustainable forest management, their effect will be malign. So far, they have been more effective in improving forest management in richer countries outside the tropics – a perverse and unintended outcome.

Criteria and indicators could be a very useful tool to define the parameters of sustainable forest management (though these could equally well be defined by guidelines), but they have become a good idea that has lost its way – the best being the enemy of the good. It is interesting to follow their history. ITTO launched the idea in a modest way; it gained speed and volume through the Helsinki and Montreal Processes, and rapidly developed a momentum of its own. The intention in ITTO was to develop a simple tool by which countries could assess the status of their forests and how well the ITTO guidelines were applied, and to report on these to ITTO.

At the first meeting of the Helsinki Process devoted to developing its criteria and indicators, the UK put forward the suggestion that the criteria should correspond to the main fields of concern, that there should be guidelines covering each of these and that reports should concern themselves with (a) whether the guidelines were complete; and (b) whether they were implemented.

But this proved too simple for the meeting. It was too simple too for the CSCE meeting in Montreal. An ideal set was designed which would completely define the nature of sustainable forest management and would be fully capable of recording progress towards this ideal. There were further regional meetings to design sets fitted to local conditions and other kinds of forest. And so they have developed the plethora of present initiatives on criteria and indicators leading to meetings to discuss equivalence and harmonization. ITTO rewrote its criteria and indicators to conform to the developing international pattern. Training courses were arranged to help nations and forest managers to complete reports based on the criteria and indicators. Standards are being developed based upon the indicators, and systems of internal auditing are being devised. There has been particular confusion between those indicators concerned with measuring the *condition of the forest resource* (something of permanent validity) with those concerned with *forest values* (the appreciation of which is transitory) and the *means of implementation* (which can be various) – a confusion between *what* and *how*.

Much the same criticism of over-elaboration might be made of the emphasis on monitoring, assessment and reporting. These are means to an end, not ends in themselves. Should we not rather concentrate resources on furthering the ends? The auditing of performance is a very useful tool to improve performance; monitoring is a useful tool to check performance against objectives; reporting imposes a useful discipline. The main objective of all of these should be to improve performance. Beyond this, they divert resources from doing so. They tend also to be used to make invidious comparisons – or for political objectives. The figures showing rates of deforestation are a case in point. They are used to fuel arguments about whether deforestation is decreasing or increasing – not about what forest should be preserved and how to ensure that it is preserved.

New priorities

The way forward lies in adopting a much more direct approach. I believe that we are in danger of becoming lost through trying to be too clever. As a result, much effort has been expended on activities which, though important, are peripheral to the main goal which might be defined at the national level, where sovereignty will continue to lie, as:

> To cherish the nation's forests and to manage them so that they make a durable contribution to national prosperity and ecological stability.

Action should instead be concentrated on changes in the use and management of forest lands.

What then should be done? What are the priorities? My choice is the following:

- End the isolation of forestry.
- Determine what forests each nation needs, provide them with legal and tenurial security and look after them.
- Be prepared to alienate other forests, especially for sustainable agriculture.
- Consider the effects of all national policies on the future of the nation's forests.
- Concentrate on those forms of forest management that are cost-effective and known to bring results, particularly protection, conservation and intensive timber production.
- Manage the remaining forests as an appreciating capital asset, extracting produce from them as market opportunities arise. Be flexible and adaptive in managing these according to their ecological characteristics and social setting.
- Respect general management guidelines but avoid being too prescriptive.
- Use criteria and indicators, auditing and monitoring to the extent that they are useful in checking and adapting management activities to conform to well-constructed guidelines.
- Develop conditions in which all forms of sustainable forest management can become financially viable – either through market price or other mechanisms. The most effective would be the market; but, if this is insufficient, services could be subsidized nationally or internationally, as appropriate.
- Devote international resources into bringing the management of tropical forests into the mainstream of national land use policies and sustainable development.
- Build national capacities so that the planning of national land use and the adaptive management of forests becomes the rule not the exception – in all forest, not only those devoted to demonstration. This requires people who have a deep understanding of the dynamics of forest ecosystems and of the social systems which they serve.
- Make the world aware of the lasting importance of forests and of wood as a raw material. Use reasoned arguments based on the facts; avoid sensationalism.
- Strengthen and streamline the role of the existing organizations, which are parties to the International Arrangement on Forests (IAF) and avoid creating any new mechanisms.
- Ensure that these existing organizations and mechanisms work in constructive harmony with one another rather than in debilitating rivalry.

This last point brings us back to the future place for ITTO. I hope that the history presented in the preceding chapters has demonstrated that ITTO has indeed been successful in forwarding the vision of its founders. This role could, perhaps, have been carried out by FAO, but governments decided otherwise: ITTO came into being. Since then, it has been influential in altering and refining the nature of the forest debate out of all proportion to its size and budget. It has also been able, within limits, to transform policy into practice. It has been a good international investment. There is no doubt that it could have been more effective; it has missed some important opportunities. But the Council is aware of these deficiencies and apparently intends to act. In doing so, it needs to be more decisive than it sometimes has been in the past. The changes introduced in the ITTA, 1994 have been beneficial – assistance directed towards the countries most in need combined with a comprehensive overview of the global timber trade.

ITTO should continue to be one of the international pillars to promote, bring about and support the wise management of tropical forest lands; but, for this, it should be given more support, particularly by consumer members and the environmental NGO community. No institution with an international role in forest management is perfect, but the international environmental NGOs appear to have migrated from one to another in search of a forum that fits their needs. It would be more constructive to take a longer view and work instead to improve the institutions they criticize. ITTO, being small, is not hindered by a cumbersome bureaucracy; it has shown its ability to adapt to changing circumstances and to alter both its policies and its practice. It has its faults; but for these to be fixed, outsiders are needed – trade and environmental NGOs in particular. Intergovernmental bodies – particularly those in the UN system – are often criticized for being bureaucratic, expensive and ineffective, yet the rapid trend towards globalization means that they will play an increasing, and important, role in shaping the world's future. New models for the way such bodies should work are, I believe, desperately needed. ITTO was first devised to address a specific concern but it has evolved into something much broader; in the process, negotiators have created something new, and the lessons to be drawn from it are many. When the ITTA comes up for renewal, there seems to be little reason to change its objectives. But the International Tropical *Timber* Organization is undoubtedly handicapped by a name that, in the view of some, carries negative overtones and certainly does not adequately capture the full scope of the Organization. Should it not, perhaps, be called the International Tropical *Forest* Organization – a name that would reflect its actual role and its future potential?

A last word: results will only be as good as the people who are striving for them. Good management of forests depends not only on sound science but also on accurate observation and sensitive judgement. Sustainable forest management will only succeed when there are many people with such talents in positions where they are able to shape the future forest landscape.

Appendix 1

Definitions[1]

Forest and related land cover

These terms have been adopted for use in the Global Forest Resources Assessment 2000. Further information on definitions has been published in FAO, 1998, *FRA 2000 Terms and Definitions*, and FAO, 2000, *On Definitions of Forest and Forest Change*, which are available on the FAO website (FRA Working Papers, 1 August 2001) at: www.fao.org/forestry/fo/fra/index.jsp.

Forest

Forest includes natural forests and forest plantations. The term is used to refer to land with a tree canopy cover of more than 10 per cent and area of more than 0.5ha.

Natural forest

A forest composed of indigenous trees, and not classified as forest plantation.

Forest plantation

A forest established by planting and/or seeding in the process of afforestation or reforestation. It consists of introduced species or, in some cases, indigenous species.

Other wooded land

Land that has either a crown cover (or equivalent stocking level) of 5 to 10 per cent of trees able to reach a height of 5m at maturity; or a crown cover (or equivalent stocking level) of more than 10 per cent of trees not able to reach a height of 5m at maturity; or a shrub or bush cover of more than 10 per cent.

Afforestation

Establishment of forest plantations on land that until then was not classified as forest. It implies a transformation from non-forest to forest.

Natural expansion of forest

Expansion of forest through natural succession on land that until then was under another land use (eg forest succession on land previously used for agriculture). It implies a transformation from non-forest to forest.

Reforestation

Establishment of forest plantations on temporarily unstocked lands that are considered forest.

Natural regeneration on forest lands

Natural succession of forest on temporarily unstocked lands that are considered forest.

Deforestation

The conversion of forest to another land use or the long-term reduction of the tree canopy cover below the minimum 10 per cent threshold.

Forest degradation

Changes within the forest that negatively affect the structure or function of the stand or site, and thereby lower its capacity to supply products and/or services.

Forest improvement

Changes within the forest that positively affect the structure or function of the stand or site, and thereby increase its capacity to supply products and/or services.

Appendix 2

The London Environmental Economics Centre Report and Certification

November 1992	The report was presented to the Council in November 1992, when it was deferred because it had only just been presented and was only available in English.
May 1993	There was great interest at the market discussions in May 1993 – a presentation by the consultants, further presentations by Brazil, Ghana, Indonesia and Malaysia, and over 40 interventions from the floor covering the whole field of the report. All the papers and a full record were then circulated to all participants, so that it might be discussed again in November.
November 1993	In November, it was considered by a joint meeting of all three Permanent Committees, but discussion was focused solely on certification. No other items in the report were discussed.

The meeting agreed that, to be workable and effective, schemes must: be acceptable to and implementable by national governments and other forest owners; be acceptable to customers in both domestic and international markets; be practical to implement, based on criteria understood at all levels of the wood market chain; be appreciative of commitment to improvement and progress towards practical achievement of sustainable forest management within a time frame agreed by the international community; and not be used as a non-tariff barrier against producer countries and should apply equally to all types of timber.

And the decision? The Executive Director was requested to engage two independent expert consultants to prepare a survey or inventory of existing and proposed certification schemes. The survey should be presented in a coherent and consistent format drawing together all essential information and indicating the objectives and mechanisms by which the scheme or proposed schemes were being or might be implemented in the markets. The survey should cover all types of timber and timber products – to be presented to the next session of Council.

May 1994	In May 1994, there was a Council Decision,[2] which registered appreciation, requested members to furnish information and, once again, requested the Executive Director to engage two consultants, this time to undertake a detailed study on markets and market segments for certified timber and timber products – to be presented to the 18th Session of Council (May 1995).

May 1995	There is no record that it was received or discussed at that session.
November 1995	In November 1995 there were two items on the agenda: (a) Study of developments in the formulation and implementation of certification schemes for all internationally traded timber and timber products; and (b) Study of markets and market segments for certified timber and timber products. Both were deferred, for presentation to Council in May 1996.
November 1995	*The Mid-Term Review* 'Although the main impetus to certification relates to timber and products in international trade, this represents only a small percentage of the total production in most tropical producer countries. To be effective as a policy instrument ... certification would need to cover all production forests. This implies a need for certification systems operating at the national level, although designed to conform with the internationally agreed principles expected to emerge eventually from the convergence of ongoing processes concerned with defining the criteria and indicators for sustainable forest management.'
May 1996	In May 1996, the two items were discussed by the Committee.[3] On the former (formulation and implementation of certification schemes), there were a number of concerns: the risk of certification discriminating against small and medium-sized forestry operations; the implication of certification for plantation forestry; the competitive advantages of plantation timber over tropical timber; the necessity of undertaking chain-of-custody certification (as opposed to forest management certification); the agencies to carry out certification and the extra costs involved; the relevance of certification to domestic markets; the appropriateness of industry-driven (ISO) or NGO-driven (FSC) schemes; and the need for ITTO also to undertake the studies of life cycle analysis of timber and substitute products. There was at best a small niche market willing to pay a green premium for certified timber. The US delegation: 'questioned the report's assumption that certification would be an acceptable policy instrument in all countries, and that the preference for certified products existed worldwide. It cautioned against the linking of national criteria and indicators for sustainable forest management (country level) with certification performance standards (management level). The US delegation also felt that any conclusions on this subject should not pre-judge on-going discussions by the IPF.' The second report (markets and market segments) was rejected as below standard! The Council passed a Decision[4] authorizing the Executive Director to engage two consultants to prepare a comprehensive update.
November 1996	The Committee was informed that the terms of reference had been drafted and the consultants retained, but a supplement was required to the original budget.
May 1997	The final report would be submitted in mid-1997.
December 1997	No record.
May 1998	The Committee noted that the final report had been circulated but no discussion had been scheduled. It was agreed that certification was of differing importance in different markets and market segments; that it was important to continue to use the ITTO forum to

discuss the topic and to allow for the sharing of experiences; and that it was important to disseminate information to the public on examples of good forest management in the tropics. It would continue its discussion at the next session.

November 1998
The Committee noted timber certification as an evolving issue requiring continuing and close monitoring by ITTO. It was informed of the Pan-European Forest Certification (PEFC) for the establishment of comparable national certification systems and their mutual recognition. It agreed to keep the issue of timber certification on its agenda.

May 1999
The Committee noted the frequent reference made during the ITTO 1999 Market Discussion on the issue of timber certification and, in particular, the proliferation of initiatives in this field. Considering the widespread and possible confusion in the marketplace over the different verification, labelling and certification schemes in various stages of consideration or implementation, the Committee requested the ITTO Secretariat to collate information on the range of schemes being developed for consideration at its [next] session.

November 1999
The Committee noted the Secretariat collation of information. It noted also the growing significance of mutual recognition within the context of timber certification and the convening of an international meeting on the potential development of a mechanism of international cooperation in forest certification and labelling … in November 1999.

May 2000
The Committee agreed to consider further developments surrounding the issue of timber certification.

November 2000
The Committee considered recent developments in timber certification that had centred on the continuing efforts to examine the possibilities and requirements for the conceptualization of an international mutual recognition framework linking the various sustainable forest management standards and certification systems now being developed… While interest in mutual recognition was growing, there was still the need for more discussion and exchange of views in order to build up confidence and common ground on the issue … ITTO would be collaborating with FAO and GTZ in hosting an international workshop at the FAO Headquarters in February 2001 that would focus on issues relating to mutual recognition which would be of particular relevance to developing countries. It was hoped that the workshop would facilitate greater awareness and understanding among developing countries concerning the ramifications and implications of the latest developments surrounding the issue of timber certification… The Committee's attention was drawn to a number of papers from the Ministerial Conference on the Protection of Forests in Europe and the Department of Agriculture, Fisheries and Forestry of Australia.

November 2000
Review of Progress towards the Year 2000 Objective:
'The initial concern that the pressure for timber certification was a front for the imposition of bans on the import of tropical timber has now abated. In fact, the consumer pressure being placed on the trade from retailers and local government authorities is having a marked effect on attitudes towards the sustainable management of temperate and boreal forests as well as tropical forests. Nevertheless, there has been sufficient effect on certain products for

a number of tropical countries to consider it important to develop their own timber certification systems. At the same time, it has begun to be recognized that the establishment of the standards of performance necessary for certification would have a beneficial effect on their own forest management. So the interest in certification is growing in countries in all three continents, and in some timber trade organizations. Certification systems are being developed and tested in a number of countries. Several of the representations that the consultants received from timber trade organizations urged ITTO to take more of the initiative in certification.'

'To balance these successes, there have also been a number of lost opportunities... ITTO was offered the opportunity to be first in the field in developing certification (then called "labelling") and it still, much to the regret of those who have submitted comments from the timber trade, seems reluctant to take the initiative in this field.'

June 2001   *Timber Advisory Group*
'ITTO to be actively involved in facilitating moves towards the mutual recognition of timber certification schemes and not endorse or be perceived to endorse any one particular scheme.'

June 2001   *Decision 10(XXX)*
Facilitate discussion. Members to submit proposals for criteria and indicators, and to build capacity for internal audit. Workshop before ITTC XXXII.

Notes

Foreword

1   Type 2 initiatives are in contra-distinction to those commitments agreed by governments through an intergovernmental process.
2   The Intergovernmental Panel on Forests (IPF) was established by the Commission on Sustainable Development (CSD) to follow up on the Statement of Principles on Forests adopted at the Earth Summit in Rio de Janeiro in 1992; it was subsequently extended as the Intergovernmental Forum on Forests. In 2000 the United Nations General Assembly institutionalized this as the United Nations Forum on Forests.

Chapter 1 The rise and fall of forests

1   Morley, 2000.
2   Where the reference is to *Homo sapiens*, 'man' and 'his' are used to include both female and male.
3   For an account of the extinction of the Megafauna in the Americas and in Australia–New Guinea, see Diamond, 1998.
4   Grove and Rackham, 2001.
5   Lane Fox, 1986.
6   Jones, 1964, quoted by Darlington, 1969.
7   Ascherson, 1996.
8   Sommer, 1976.
9   The origin of this idea is described in Chapter 6. It has, at different times, been known as Target 2000, the Year 2000 Objective and, most recently, as Objective 2000.

Chapter 2 Sustainable forest management: a response to destruction

1   This chapter draws on material prepared for IUCN in the 1970s (acknowledged with appreciation) and on sections of *No Timber without Trees* (Poore et al, 1989).
2   FAO, 2001a.
3   For example, at least 11 quite distinct forest types were found between 600 and 3300m altitude on the southwest slopes of Mount Kinabalu in Sabah, Malaysia. Ho and Poore, 1968.
4   Some countries have mapped their vegetation using classifications that take some account of the floristic and faunistic composition of their forests. National

forest inventories are also valuable in providing information on the presence and abundance of tree species.

5   See the internationally accepted definitions of 'forest' and 'deforestation' in Appendix 1 (p261).

6   See Geist and Lambin, 2002.

7   Poore, 1976. These were revised and republished in Poore and Sayer, 1987.

8   Poore, 1974.

9   Concern about the future of the tropical forest was still largely confined to scientists and professionals. The public was still unaware. An attempt by the then World Wildlife Fund (International) to run a fund-raising campaign on the tropical forest in the late 1970s had to be discontinued for lack of support.

10   Many historical studies in temperate countries have shown that forests were best managed in the past when there was a clear economic objective in maintaining production from them.

11   This division into two was eventually to be paralleled by the two levels of guidelines (or criteria and indicators) – the 'national level' and the 'forest management unit level'.

12   IUCN, UNEP, WWF, 1980.

13   Another extreme view is that a forest managed for timber production should retain the full biodiversity of unexploited forest. On this view no management for timber would be considered sustainable.

14   ITTO studies showed that fires were more intense and the damage greater in forest which had been logged.

15   It may be necessary to modify the objective to meet changing circumstances; the decision to do so should be explicit.

16   Most notably since UNCED in 1992.

17   A useful grouping of producer countries is reproduced in Table 16.1.

18   Burns, 1986; HIID, 1988; Repetto and Gillis, 1988.

19   Kaimowitz, personal communication by email.

20   For a very full treatment of these issues, see Mayers and Bass, 1999.

Chapter 3 Genesis of a treaty: the ITTA takes shape

1   A chronology of events is given in Table 3.1.

2   The ITC was established on 1 January 1968 in pursuance of a General Assembly Resolution and the decision of the Contracting parties to the General Agreement on Tariffs and Trade (GATT).

3   The words 'ambitious and complex' were contained in an internal FAO memorandum from A Lacayo to Om P Mathur and refer to a note 'Elements significant to the conceptual development of the TTB' addressed by FAO to UNDP in January 1973.

4   It would not have been impossible at this stage to design a mechanism whereby the work was distributed between FAO and UNCTAD, one of which had primary responsibility for forestry and the other for trade. One can only speculate why this was not done.

5   ITC/POP/89Rev.1, March 1974.

6   Pleydell, Geoffrey. Written communication.

7   Three different draft agreements are mentioned in this chapter:

     1   The draft prepared by the ITC and adopted in October–November 1977. This agreement was not signed and therefore lapsed.

2   The discussion draft prepared for the Workshop of Producer Countries in Abidjan.

3   The draft, prepared by Japan for UNCTAD, which was the basis for the ITTA, 1983.

8   In May 1975, 15 African countries, meeting in Bangui in the Central African Republic, took their own initiative, by signing an agreement which led in due course to the African Timber Organization (ATO).

9   This was based on a paper *Consideration of International Measures for Tropical Timber* (TD/B/IPC/TIMBER/2).

10   For example: *A Preliminary Report on the Status of International Financial Assistance Relevant to Reforestation and Forest Management* (TD/B/IPC/TIMBER/27 and addenda); *The Present Status of Tropical Forest Resources; A Preliminary Appraisal* (TD/B/IPC/TIMBER/28); and *Research and Development: Identification of Priorities for International Action on Forest Management and Reforestation* (TD/B/IPC/TIMBER/29).

11   Afforestation using fast growing broad-leaved and coniferous species was mentioned but was never pursued in depth.

12   A Leslie, written communication.

13   'Should the Agreement not have entered into force by 31 March, 1985, the Secretary General of the United Nations is requested to convene a meeting of those countries which have signed the Agreement and deposited an appropriate instrument, to decide whether to put the Agreement into force among themselves. As only 2 producer countries would be invited to such a meeting, it was considered that the consumer countries would consider such a meeting inopportune.' Cording, 1985.

14   Two weeks earlier, Mr McIntyre, the Deputy Secretary General of UNCTAD had invited the Ambassadors of tropical timber producing countries with Missions in Geneva to a meeting in order to urge most rapid action from producing countries. At their request, he sent an urgent cable to all 36 timber producing countries.

Chapter 4 ITTO's early days: optimism and experiment

1   WRI, World Bank, UNDP, 1985.

2   FAO, 1985. The Tropical Forestry Action Plan reviewed five fields: forestry in land use; forest-based industrial development; fuelwood and energy; conservation of tropical forest ecosystems; and institutions. It was designed to identify areas in which additional investment or technical assistance were required. In 1987, the four sponsors, WRI, World Bank, UNDP and FAO, convened the Bellagio Conference on Tropical Forests to gain political support for the Plan and issued a summarized version of it.

3   IIED and GOI, 1985. See also Poore, Ross and Setyono, 1985.

4   Quotations from Council sessions given in this book are taken from the reports of the sessions published by the Secretariat; in most cases the quotes are paraphrases of what was actually said.

5   These were in fact guidelines for ITTO action in this field.

Chapter 5 First assessment: living in a fool's period

1   A forest composed of indigenous trees, and not classified as forest plantation. (See Appendix 1).

2   IIED, 1988 and Poore et al, 1989. This chapter draws extensively on *No Timber without Trees*. The conclusions presented to the ITTC were partly rewritten for the book – to make them more accessible, and to correct certain misunderstandings. Many who read the report did not realize that substantial areas of forest were *nearly* being managed sustainably; only one or two of the preconditions were lacking. The situation was not necessarily so desperate as some commentators made out.

3   Country studies were carried out by Simon Rietbergen for Africa; by Peter Burgess for Asia; by Timothy Synnott in Latin America and the Caribbean and by Duncan Poore in Queensland. An independent commentary on the subject was commissioned from John Palmer based on his extensive field experience and knowledge of the literature. Both the Forest Resources Division of FAO and the Coordinating Unit of the TFAP were consulted. If there was known to be important experience in non-member countries, this was taken into account. The study was coordinated for IIED by Duncan Poore.

4   In 1980 the estimated total area of closed broad-leaved tropical forest was 1200 million hectares of which 860 million were 'productive'. It was estimated that this area of productive forest would be reduced to 828 million by 1985. (Lanly, 1988. Written communication.)

5   This is also true, of course, of illegal logging.

6   All of these depend, of course, upon adequate resources for finance and personnel.

7   Harvard Institute of International Development and Japanese Overseas Forestry Consultants Association.

Chapter 6 From Abidjan to Bali: a radical new agenda

1   World Commission on Environment and Development, 1987.

2   The Brundtland definition of sustainable development strongly echoed the definition of conservation in the much earlier *World Conservation Strategy,* in which the term 'sustainable development' was first launched on the world scene.

3   Panayotou and Ashton, 1992.

4   Quotations in this section that are unattributed in the main text are taken from the Draft Report of the 6th Session of the ITTC.

5   This Mission was not a success. The Delegation had miscalculated the degree of knowledge of their critics and the sophistication of their attitudes. The material presented appeared condescending to its audiences and, in some instances, led to a hardening of attitudes.

6   ITTC Resolution 1(VI).

7   For those who take a holistic view of ecology – that it should include the study of the human species – this would be automatically included by Article 1(h) of the ITTA, 1983: 'maintaining the ecological balance.'

8   This and succeeding paragraphs are drawn from the Draft Report of the 7th Session of the ITTC.

9   Quotation from Pleydell. Written communication.
10   The following quotations are drawn from the Draft Report of the 8th Session of the ITTC.

Chapter 7 The case of Sarawak

1   There was also a body of opinion that thought that *all* the recommendations should be included in the final section and that the whole range of values for sustainable harvest should be presented, specifying the conditions needed for each to be attained.
2   The Federal Government also coordinates policy through the National Forestry Council which includes the Chief Ministers of all States.
3   A history of successive land laws was provided for the Mission by the Land and Survey Department. An independent published account is found in Hong, 1987, ch. 4.
4   Quoted by Hong, and kindly provided for the Mission by Sahabat Alam Malaysia (SAM).
5   The Mission received from the Land and Survey Department a xerox copy of the Land Code incorporating all amendments and modifications up to 6 December 1988.
6   It should be noted that the Sarawak definition of PFE excludes TPAs whereas the ITTO definition includes them. The Sarawak 'forests on State land' correspond broadly to the 'conversion forests' in the ITTO definition.
7   From information in published reports, plans and interviews, and the statistics of the Forest Department and other agencies.
8   Approximately 1 million hectares for TPAs, 3.5 million hectares for permanent agriculture, 2 million hectares for additional cultivation and a small area for urban development.
9   The figures for swamp were based on FAO data for area and yields; the figures for mangrove were derived from yields outside Sarawak.
10   One area of 4000ha sampled showed only 45 trees per ha which would qualify for future crops.
11   Forest Department statistics.
12   From the Draft Report of the 8th Session of the ITTC.
13   Similarly, the NGOs discounted the TFAP rather than building upon its potential. The relations between ITTO and the environmental NGOs have been examined in detail by Gale (1996). This will be considered more fully in Chapter 18.
14   Kavanagh et al, 1989.
15   Quotations in this section that are unattributed in the main text are taken from the Draft Report of the 9th Session of the ITTC.
16   The revenues from oil accrue to the Federal Government.
17   Centeno in the Draft Report of the 9th Session of the ITTC.
18   Quotations in this section that are unattributed in the main text are taken from the Draft Report of the 11th Session of the ITTC.
19   It is perhaps significant that there was no mention that Sarawak might receive a higher proportion of Federal oil revenues, even though these were mainly derived from sources in Sarawak.
20   As paraphrased in the Draft Report of the 19th Session of the ITTC.

Chapter 8 Ferment 1990–1992

1   One bizarre feature of this sensitivity was that it became unacceptable in international meetings to speak of 'indigenous peoples' but only of 'indigenous people', thus denying any individuality to distinctive groups of people.

2   Dudley, 1992.

3   The TFAP-CG was to have 36 members: 14 developing countries, 7 developed countries, UNEP, the Forestry Forum for Developing Countries (FFDC), the TFAP Advisors Group and NGOs.

4   An excellent general account of the post-UNCED international developments is given in a sequence of publications from the Commonwealth Forestry Institute (Grayson, 1993; Söderlund and Pottinger, 2001); and the text of the international conventions are collected in Schmithüsen and Ponce-Nava, 1996.

5   Dr Maini. Written communication.

6   ITTO, 1990c.

7   Oxford Forestry Institute, 1991.

8   Executive Director in ITTC(X)/12.Rev.1

9   ITTC(X)/12.Rev.1

10 Throughout the record of these discussions the term 'Sustainability 2000' is used. This is synonymous with the 'Year 2000 Objective'.

11 The items specifically related to identifying and reporting on progress towards the Year 2000 Objective are italicized (author's emphasis). The Decision was practically all-embracing, including many measures for the improvement both of forest management and of market mechanisms. In addition, some items were included which became a regular feature of later international negotiation. These included: liberalized and fairer trade, the transfer of technology, and the assessment of resources considered necessary to attain the Year 2000 Objective.

12 Quotations in this section that are unattributed in the main text are taken from the Draft Report of the 10th Session of the ITTC.

13 Author's emphasis.

14 The ITTC decided subsequently that these three 'project ideas' should be converted into ITTO 'activities' in order to accelerate implementation.

15 Taken from the Draft Report of the 11th Session of the ITTC.

16 Quotations in this and succeeding paragraphs that are unattributed in the main text are taken from the Draft Report of the 12th Session of the ITTC.

17 Quotations in this and succeeding paragraphs that are unattributed in the main text are taken from the Draft Report of the 13th Session of the ITTC.

18 ITTC Decision 6(XV).

Chapter 9 Tropical forests, or all forests? Renegotiating the ITTA

1   The following account is based on an internal document prepared by Koy Thompson for IIED.

2   Chapter 11 was, rather unfortunately, given the title 'Combating deforestation' rather than a more positive title, which would stress the wise use and management of forests. This has meant that much subsequent discussion has been overly focused on the extent of forest cover rather than its quality.

3   Informal Working Group on ITTA Renegotiation, Report by the Chairman. ITTC PrepCom (I)/3.
4   This was the term most generally used at the time to describe the Year 2000 Objective; it will be retained, where appropriate, in this chapter.
5   Quotations in this section that are unattributed in the main text are taken from ITTC PrepCom (I)/3.
6   Quotations in this section that are unattributed in the main text are taken from ITTC PrepCom (II).
7   This is an interesting change of attitude from that held by the producers during the negotiations in 1976.
8   Emphasis in the speaker's text.
9   Quotations in this section that are unattributed in the main text are taken from the Draft Report of the 14th Session of the ITTC.
10 The LEEC report was the principal topic in the market discussions at this session of Council. Its findings will be discussed in the next chapter.
11 Quotations in this section that are unattributed in the main text are taken from the Draft Report of the 15th Session of the ITTC.
12 At the request of the Delegate of the People's Republic of China, the word 'successfully' was deleted before 'concluded'.
13 Interview with Andrew Bennett, *Tropical Forest Update* 5/3, September 1995.

Chapter 10 Has the tropical timber trade any leverage? Policies 1991–1995

1   This last study became lost at the Council level: in ITTC(XV) p71 it was reported as 'incomplete, therefore deferred'; in ITTC(XVI) p65 the Committee on Economic Information and Market Intelligence was given a detailed presentation; by ITTC(XVII) p55 'final reports were still being prepared; by ITTC(XVIII) p70 'report finalized, circulated and being considered by the Government of Ghana.' Afterwards, nothing!
2   Mok and Poore, 1991.
3   This is different from the FAO definition used in FRA 2000 (see Appendix 1).
4   The definition accepted by Council has stood the test of time, although there are some who would wish the 'unacceptable impacts' to be spelt out unequivocally.
5   Annexes 1–3 in Mok and Poore, 1991.
6   Annexes 4 and 5 in Mok and Poore list the recommended actions to deal with Categories 2 and 3 and Annex 6 contains the checklist (see Box 10.1)
7   ITTC Decision 3(XII).
8   A project for dissemination and training in the ITTO Guidelines and Criteria was approved and financed in November 1992 (PD 39/92 Rev.1 (F)).
9   Annex to Decision 5(XIII) which was finally approved in the 14th Session.
10 This title proved to be unfortunate and led to misunderstandings, 'Accounting' was taken to mean economic or financial accounting rather than giving an account of the resource; and the use of the acronym FRA led to confusion with FAO's Forest Resource Assessment (FRA).
11 IIED/WCMC, 1994.
12 A cynical observer might be justified in thinking that the whole process of developing criteria and indicators was used by some as a diversionary tactic – a way of deferring the actual improvement of management in the field.

13   The full arguments can be found in LEEC, 1992 and the book which was based
      upon it, Barbier et al, 1994. This account is based on these two sources.
14   Although 70 per cent of plywood was exported, this was mainly from Asian
      producers and the pattern was not reflected in other regions.
15   From LEEC, 1992.

Chapter 11 The many roads from UNCED

1   Grayson, 1993; Grayson and Maynard, 1997; Söderlund and Pottinger, 2001.
2   Ministry of Agriculture and Forestry, Helsinki, 1993. A First Ministerial
      Conference on the Protection of Forests in Europe was held before Rio, in
      1990.
3   The last phrase is especially significant, because it recognizes the fact that some
      non-forest ecosystems may have a value (eg for biodiversity) that is greater than
      the forest ecosystems with which they might be replaced.
4   CSCE, 1994.
5   See the papers by Maini and by Poore in CSCE, 1994.
6   CIFOR, 1995.
7   National Forest Programmes were defined in 1997 by the International Panel of
      Forests and, after that, the acronym NFP came into general use.
8   Ball, J. Written communication.
9   For further details about the Forest Stewardship Council and certification,
      consult Upton and Bass, 1995.
10   Because of the ambiguities surrounding the use of the term 'sustainable
      management', the FSC prefers to use 'well managed'.
11   This summary is based on the very full account given in the Commonwealth
      Forestry Association publication *The World's Forests: Rio+8* and on documents of
      ECOSOC and the CSD.
12   E/CN.17/1997/12 20 March 1997.
13   A possible exception is the all-important question of the place of forests in the
      national balance of uses of the land.
14   E/CN.17/2000/14 20 March 2000.

Chapter 12 ITTO's road to 1995: ENGOs diverge

1   Quotations in this section that are unattributed in the main text are taken from
      the Draft Report of the 15th Session of the ITTC.
2   The numbers in the WWF delegation (including representatives of national
      branches) from the 1st to the 20th Sessions of ITTC were as follows: +, +, 5, 9,
      7, 7, 12, 13, 7, 10, 11, 8, 15, 18, 5, 3, 1, 0, 1, 0 (present but numbers not recorded
      at the first two sessions).
3   Quotations in this and succeeding paragraphs that are unattributed are taken
      from the Draft Report of the 16th Session of the ITTC.
4   This is probably an unreasonable suspicion.

5   The quality of management of the peat swamp forest was considered good by the Sarawak Mission with the exception of the management of 'ramin'. If this management were to be continued, it would lead to reduced yield in future rotations but would be unlikely to lead to extinction. However, the export of illegally felled 'ramin' was also an issue.

6   Compare the arguments that are advanced in favour of listing species in Appendix III (see Chapter 16). See also the explanation of the three appendices below. The error is to apply a global solution (Appendix II) to a problem that is local.

7   ITTC Decision 3(XVI).

8   ITTC Decision 3(XVII).

9   Taken from the Draft Report of the 18th Session of the ITTC.

10 ITTC Decision 5(XVIII).

11 It is interesting that international reporting on trade statistics has proved easier to organize, although even in this field there can be difficulties associated with trade in timber which is illegally acquired.

12 Quotations in this and following sections that are unattributed in the main text are taken from the Draft Report of the 19th Session of the ITTC.

13 ITTC Decision 2(XVIII).

14 Kemp and Phantumvanit, 1995. ITTC(XIX)/6.

15 The contribution made by ITTO projects will be reviewed in Chapter 15.

16 ITTC Decision 3(X) reaffirmed in November 1991 in ITTC Decision 7(XI)

17 ITTC Decision 2(XII)

18 ITTC Decision 3(XIII)

19 ITTC Decision 3(XIV)

20 ITTC Decision 2(XVIII)

21 ITTC(XIX)/5 to which is appended a working paper prepared by AJ Leslie and Mauro S Reis.

22 Mr Mankin is almost the only member of an international environmental NGO to continue to attend ITTO meetings regularly.

Chapter 13 1995–2000: getting on with the job

1   The final version of this appeared in November 2000. ITTC(XIII)/9/Rev. 2. Poore and Thang, 2000. This incorporated some country reports which were submitted after the first version was prepared.

2   The Forest Law, 1996 was not enacted until the Mission was nearly over.

3   ITTC(XXI)/9, 1996.

4   Taken from the Draft Report of the 20th Session of the ITTC.

5   Personal communications at the AIMA Meeting held in Quito, 29 January to 2 February 2001.

6   An account of the various international initiatives on criteria and indicators is given in FAO, 2001b, 2001c. They are (February, 2002): African Timber Organization; the Dry-Zone Africa Process; ITTO; Lepaterique Process of Central America; the Montreal Process (temperate and boreal forests); the Pan-European Forest Process; and the Tarapoto Process (Amazon forest). Some members of ITTO also subscribe to the ATO, Lepaterique processes and Tarapoto proposal.

7   ITTO, 1998a.

8   ITTO, 1999a.
9   The 'format' has not been published but is made available in either hard copy or electronic form.
10   Forestry Agency, Tokyo, 2001 (see FAO, 2001c).
11   ITTO, 1998b.

Chapter 14 Year 2000 Report: the curate's egg

1   The full report is ITTC(XXXVIII)/9/Rev.2.
2   Country reports were received from Australia, Bolivia, Brazil, Cambodia, Cameroon, Canada, China, Colombia, Congo, Ecuador, European Union, Gabon, Ghana, Guyana, Indonesia, Japan, Liberia, Malaysia, Myanmar, Nepal, New Zealand, Norway, Panama, Papua New Guinea, Peru, Philippines, Republic of Korea, Sweden, Switzerland, Surinam, Thailand, Togo, Trinidad & Tobago, Venezuela, the UK and the US.
3   Timber does not, of course, come only from a country's PFE nor only from natural forest. It may come from plantations, farm forestry, the clearing of land for permanent settlement or secondary forest. It may be harvested either legally or illegally.
4   The detailed description of the situation in each country is not included.
5   The dramatic decline in production for Thailand and the Philippines is due to their logging ban policy. Except for Malaysia and Myanmar, and to a lesser extent for India, the PFE area estimates in this table are highly indicative. In India, for example, a significant proportion of its recorded Forest Reserves is not forested.
6   Also the increased demand for timber for domestic and commercial uses.
7   Called 'international aspects' in the format for reporting.
8   Though far less than estimated requirements!
9   But see the report in Chapter 16.

Chapter 15 Reaction: false start, new energy

1   Decisions on the extension of the Agreement until 31 December 2003; ex-post evaluation (not before time!), a framework for auditing systems and the participation of civil society.
2   ITTC Statement 1(XXVIII).
3   ITTC(XXVIII)/29.
4   Quotations in this section that are unattributed in the main text are taken from the Draft Report of the 28th Session of the ITTC.
5   Wendy Baer in ITTC(XXVIII)/29.
6   Many of these recommendations involve partnerships between producer countries, consumer countries and ITTO. They have therefore not been classified into sections directed separately at producers, consumers and ITTO.
7   Key phrases are in italic type (author's selection).
8   Quotations in this section that are unattributed in the main text are taken from the Draft Report of the 29th Session of the ITTC.
9   ITTC Decision 2(XXXI).
10   ITTC(XXXI)/7, 3 August 2001, includes the Report of the Expert Panel on the New ITTO Action Plan and, as Annex F, Working Paper for the New ITTO Action Plan (July 2001) by Ivan Tomaselli and Patrick Hardcastle.

11 The contribution made by ITTO projects to sustainable forest management is the subject of Chapter 16.
12 ITTC(XXXI)/10, September 2001, in which fuller references are given.
13 It is also discouraging that most of the issues identified by this Mission have been raised by successive missions to Indonesia since 1980 without any apparent effect.
14 ITTC Decision 6(XXXI).
15 ITTC Decision 10(XXX).
16 Quotations in this section that are unattributed in the main text are taken from the Draft Report of the 31st Session of the ITTC.

Chapter 16 Policies into action: some case studies

1 For example, all projects above a certain size must provide annual independent audits and six-monthly technical and financial progress reports.
2 ITTO, 1998b.
3 ITTC Decision 3(XII).
4 Estimates produced for the Year 2000 study.
5 See UNESCO, 2002 for a review of the Biosphere Reserve concept.
6 Sayer, 1991.

Chapter 17 ITTO: agent of change

1 There have only been three occasions where a vote has proved necessary: on the location of the headquarters, the appointment of the first Executive Director and, once, on a budget issue.
2 It is instructive to compare the estimate of US$2200 million per annum required to implement Objective 2000 in all producer countries with the estimate of the cost of illegal logging to the Indonesian economy of US$2000 million per annum!
3 It is unfortunate that the very effective methods of keeping carbon out of atmospheric circulation through preserving old growth forests and deep peat deposits are not parts of the Clean Development Mechanism.
4 Country-based NGOs continued to attend sessions of the ITTC, especially when they occurred in their continent.
5 Some aspects of this dual role of NGOs have already been touched upon in Chapter 7 in relation to WWF. Gale (1996) made a detailed examination of ITTO and in particular the Sarawak Mission. He argued that 'the lack of progress in balancing tropical forest utilization and conservation [resulted] from a tacit alliance between producing and consuming country governments and the tropical timber industry to block the negotiation of the norms, procedures and compliance mechanisms to instantiate a genuinely sustainable Tropical Timber Trade Regime'. He concluded: 'Further intensive, theoretical and practical political action by the environmental movement is required to break this government-industry alliance. Environmentalists must continue to deconstruct and reconstruct the three key sets of ideas that make the alliance possible – "sustainable forest management", modern economic theory and unqualified state sovereignty.' Others may disagree!

Chapter 18 Changing landscapes, future prospects

1   This last chapter is a personal perspective.
2   These arguments do not apply to the preservation or potential extinction of wide-ranging or migratory animals. Different measures must be designed for their preservation over and above those appropriate for forest types.
3   See, eg, Mueller-Dombois and Ellenberg, 1974.
4   See Poore, 1963, and 1968; Ho, Newbery and Poore, 1987.
5   Whitmore and Sayer, 1992.
6   The term 'food miles' is now entering the vocabulary. Perhaps we should also think of 'wood miles'.
7   Sobral, 2001.
8   The idea of the 'conservation concession' introduced by Conservation International is promising, but will have the effect of increasing protected areas rather than paying for the services provided by sustainable management for timber.

Appendices

1   Source: FAO, 2001a.
2   Decision 2(XVI).
3   ITTC(XX)/19.
4   Decision 4(XX).

References

Much of the material in this book was derived from the official papers of ITTO, especially Council documents. These are prefixed ITTC with the number of the Council session to which they refer bracketed in roman numerals – eg ITTC(XI)/5.

Anon (1992) *Model for a Convention for the Conservation and Wise Use of Forests*, GLOBE International and AIDEnvironment.

Ascherson, Neal (1996) *Black Sea: The Birthplace of Civilisation and Barbarism*, Vintage, London.

Barbier, Edward B, Joanne Burgess, Joshua Bishop and Bruce Aylward (1994) *The Economics of the Tropical Timber Trade*, Earthscan, London.

Burns, D (1986) *Runway and Treadmill Deforestation*, IUCN/IIED Tropical Forestry Paper No 2, London.

CIFOR (1995) *Forest Research: A Way Forward to Sustainable Development*, CIFOR, Bogor, Indonesia

Cording, Ulrich (1985) in Proceedings of the IIED Seminar on the International Tropical Timber Agreement, 8–10 March, unpublished MS.

Cranbrook, Lord (1990) *The Promotion of Sustainable Forest Management in Sarawak: A Case Study. A Shortened Version*, Royal Geographical Society, London.

CSCE (1994) *Seminar of Experts on Sustainable Development of Boreal and Temperate Forests* (Technical Report and Annex 1), Natural Resources Canada, Canadian Forest Service, Ottawa.

Darlington, C D (1969) *The Evolution of Man and Society*, George Allen and Unwin, London.

DENR (2000) 'Forest Statistics', Forest Management Bureau, Department of Environment and Natural Resources, Republic of the Philippines, website: http://www.denr.gov.ph

Diamond, Jared (1998) *Guns, Germs and Steel*, Vintage, London.

Dudley, N (1992) *Forests in Trouble: A Review of the Status of Temperate Forests Worldwide*, WWF, Gland.

FAO (1985) *Tropical Forestry Action Plan*, Committee on Forest Development in the Tropics, FAO, Rome.

FAO (1998) *FRA 2000 Terms and Definitions*, FRA Working Paper, 1 August 2001, available on the FAO website at: www.fao.org/forestry/fo/fra/index.jsp.

FAO (1999) *State of the World's Forests*, FAO, Rome.

FAO (2000) *On Definitions of Forest and Forest Change*, FRA Working Paper, 1 August 2001, available on the FAO website at: www.fao.org/forestry/fo/fra/index.jsp.

FAO (2001a) *State of the World's Forests 2001*, FAO, Rome.

FAO (2001b) *Criteria and Indicators for Sustainable Forest management: A Compendium*, Forest Management Working Paper FM/5. FAO, Rome.

FAO (2001c) 'Use of Criteria and Indicators for Monitoring, Assessment and Reporting on Progress toward Sustainable Forest Management in the United

Nations Forum on Forests' (MAR Yokohama SFM/02), in Forestry Agency, Tokyo, *Proceedings of the International Expert Meeting on Monitoring, Assessment and Reporting on the Progress toward Sustainable Forest Management*, 5–8 November, 2001, Yokohama, Japan.

Gale, Fred P (1996) 'The ecological political economy of global environmental cooperation: A case study of the International Tropical Timber Organization in the making of the tropical timber trade regime', PhD thesis, Carleton University, Ottawa.

Geist, Helmut J and Eric F Lambin (2002) 'Proximate Causes and Underlying Driving Forces of Tropical Deforestation', *Bioscience* **52**(2): 143–150.

Grayson, A J (ed.) (1993) *The World's Forests: International Initiatives since Rio*, Commonwealth Forestry Association, Oxford.

Grayson, A J and W B Maynard (1997) *The World's Forests – Rio+5: International Initiatives towards Sustainable Forest Management*, Commonwealth Forestry Association, Oxford.

Grove, A T and Oliver Rackham (2001) *The Nature of Mediterranean Europe: An Ecological History*, Yale University Press, New Haven and London.

HIID (1988) *The Case for Multiple Use Management of Tropical Hardwood Forests*. Study prepared for ITTO by the Harvard Institute for International Development.

Ho Coy Choke and M E D Poore (1968) 'The Value of Mount Kinabalu National Park, Malaysia, to Plant Ecology', in Talbot, L M and H Talbot (eds), *Conservation in Tropical South-East Asia*, IUCN Publications New Series No 10. IUCN, Morges, Switzerland.

Ho, C C, D McC Newbery and M E D Poore (1987) 'Forest Composition and Inferred Dynamics in Jengka Forest Reserve, Malaysia', *Journal of Tropical Ecology* **3**: 25–56.

Hong, Evelyn (1987) *Natives of Sarawak: Survival in Borneo's Vanishing Forests*, Institut Masyarakat, Malaysia.

IIED (1988) *Pre-Project Report Natural Forest Management for Sustainable Timber Production*, 5 vols. PPR 11/88 (F).

IIED/Government of Indonesia (1985) *Forest Policies in Indonesia: the Sustainable Development of Forest Lands*, IIED, London.

IIED Forestry and Land Use Programme/WCMC (1994) *Forest Resource Accounting: Stock-Taking for Sustainable Forest Management*, IIED, London.

International Tropical Timber Agreement, 1983, United Nations, New York, 1984.

International Tropical Timber Agreement, 1994, United Nations, New York and Geneva, 1994.

ITTO (1990a) *Guidelines for Sustainable Management of Natural Tropical Forests*, ITTO Policy Development Series No 1.

ITTO (1990b) *The Promotion of Sustainable Forest Management in Sarawak: A Case Study*, Report submitted to the ITTC by Mission established pursuant to Resolution I(VI). ITTC(VIII)/7.

ITTO (1990c) *Criteria and Priority Areas for Programme Development and Project Work. ITTO Action Plan*, ITTC(IX)/6 Rev.1.

ITTO (1996) *Annual Review and Assessment of the World Timber Situation 1995*, ITTO, Yokohama.

ITTO (1998a) *Criteria and Indicators for the Sustainable Management of Natural Tropical Forests*, ITTO Policy Development Series No 7.

ITTO (1998b) *ITTO Libreville Action Plan 1998–2001*, ITTO Policy Development Series No 8.

ITTO (1999a) *Manual for the Application of Criteria and Indicators for Sustainable Management of Natural Tropical Forests*. Part A/National Indicators and Part B/Forest Management Unit Indicators, ITTO Policy Development Series Nos 9 and 10.

ITTO (1999b) *Annual Review and Assessment of the World Timber Situation 1998*, Yokohama.

IUCN, UNEP, WWF (1980) *World Conservation Strategy*, Gland, Switzerland.

Jones, Gwyn (1964) *The Norse Atlantic Saga*, Oxford University Press, Oxford

Kavanagh, Mikkaail, Abullan Abdul Rahim and Christopher J Hails (1989) *Rainforest Conservation in Sarawak: An International Policy for WWF*, WWF Malaysia, Kuala Lumpur and WWF International, Gland.

Kemp, Ronald H and Dhira Phantumvanit (1995) *1995 Mid-Term Review of Progress Towards the Achievement of the Year 2000 Objective*, ITTC(XCIX)/6.

Lane Fox, Robin (1986) *Alexander the Great*, Penguin, London.

LEEC (1992) *The Economic Linkages between the International Trade in Tropical Timber and the Sustainable Management of Tropical Forests*, Main Report, IIED, London.

Maini, J S (1994) 'Sustainable Development of Forests: A Systematic Approach to Defining Criteria, Guidelines, and Indicators', in CSCE *Seminar of Experts on Sustainable Development of Boreal and Temperate Forests*, Natural Resources Canada, Canadian Forest Service, Ottawa.

Mayers, James and Stephen Bass (1999) *Policy That Works for Forests and People*, Policy Series No 7: Series Overview, IIED, London.

Ministry of Agriculture and Forestry, Helsinki (1993) Ministerial Conference on the Protection of Forests in Europe, 16–17 June 1993 in Helsinki: Documents.

Ministry of Environment and Forests, Government of India (1999) *National Forestry Action Programme – India*, Government of India.

Mok, S T and Duncan Poore (1991) *Criteria for Sustainable Tropical Forest Management*, report prepared for ITTO. ITTC(XI)/6 Appendix 1.

Morley, Robert J (2000) *Origin and Evolution of Tropical Rain Forests*, John Wiley, New York.

Mueller-Dombois, Dieter and Heinz Ellenberg (1974) *Aims and Methods of Vegetation Ecology*, John Wiley, New York, London.

Oxford Forestry Institute (1991) *Incentives in Producer and Consumer Countries to Promote Sustainable Development in Tropical Forests*, ITTO Pre-project report.

Panayotou, Theodore and Peter S Ashton (1992) *Not by Timber Alone: Economics and Ecology for Sustaining Tropical Forests*, Island Press, Washington.

Poore, Duncan (1963) 'Problems in the Classification of Tropical Rain Forest', *Journal of Tropical Geography*, **17**: 12–19.

Poore, Duncan (1968) 'Studies in Malaysian Rain Forest: I. The Forests on Triassic Sediments in Jengka Forest Reserve', *Journal of Ecology*, **56**: 143–196.

Poore, Duncan (1974) 'The values of tropical moist forest ecosystems and the environmental consequences of their removal', position paper prepared for the FAO Technical Conference on the Tropical Moist Forests.

Poore, Duncan (1976) *Ecological Guidelines for Development in Tropical Rain Forests*. IUCN, Morges, Switzerland.

Poore, Duncan (1994) 'Criteria for the Sustainable Development of Forests', in CSCE *Seminar of Experts on Sustainable Development of Boreal and Temperate Forests*, Natural Resources Canada, Canadian Forest Service, Ottawa.

Poore, Duncan and Jeffrey Sayer (1987) *The Management of Tropical Moist Forest Lands: Ecological Guidelines*, IUCN, Gland, Switzerland and Cambridge, UK.

Poore, Duncan and Thang Hooi Chiew (2000) *Review of Progress towards the Year 2000 Objective*, ITTC(XXVIII)/9/Rev.2.

Poore, Duncan, Peter Burgess, John Palmer, Simon Rietbergen and Timothy Synnott (1989) *No Timber without Trees: Sustainability in the Tropical Forest*, Earthscan, London.

Poore, Duncan, Michael Ross and Setyono Sastrosumarto (1985) 'A review of policies affecting forest lands in Indonesia', paper delivered at the 9th World Forestry Congress, Mexico City.

Repetto, Robert and Malcolm Gillis (eds) (1988) *Public Policies and the Misuse of Forest Resources*, Cambridge University Press, Cambridge.

Sayer, Jeffrey (1991) *Rainforest Buffer Zones: Guidelines for Protected Area Managers*, IUCN, Gland, Switzerland.

Schmithüsen, Franz and Diana Ponce-Nava (1996) *A Selection of Texts of International Conventions and Other Legal Instruments Relating to Forests*, Grundlagen und Materialien Nr 96/2, Departement Wald- und Holzforschung, ETH Zürich.

Sobral Filho, Manoel (2001) 'The Tropical Forest Dilemma', 3rd Jack Westoby Lecture, Canberra, 9 August 2001, MS.

Söderlund, Mia and Alan Pottinger (eds) (2001) *Rio+8: Policy, Practice and Progress towards Sustainable Management*, Commonwealth Forestry Association, Oxford.

Sommer, A (1976) 'Attempt at an Assessment of the World's Tropical Forest Cover', *Unasylva* **28** (2/3).

Stringer, Chris and Robin McKie (1996) *African Exodus: The Origins of Modern Humanity*, Random House, London.

Tomaselli, Ivan and Patrick Hardcastle (2001) Working Paper for the New ITTO Action Plan. Annex F to ITTC(XXXI)/7.

UNCED (1992) *Agenda 21*.

UNCED (1992) *Non-legally Binding Authoritative Statement of Principles for a Global Consensus on the Management, Conservation and Development of All Types of Forests*. (Forest Principles).

UNESCO (2002) *Biosphere Reserves: Special Places for People and Nature*, UNESCO, Paris.

Upton, Christopher and Stephen Bass (1995) *The Forest Certification Handbook*, Earthscan, London.

Whitmore, T C and J A Sayer (eds) (1992) *Tropical Deforestation and Species Extinction*, Chapman and Hall, London, New York, Tokyo, Melbourne, Madras and IUCN, Gland.

World Commission on Environment and Development (1987) *Our Common Future*, Oxford University Press, Oxford.

WRI, World Bank, UNDP (1985) *Tropical Forests: A Call for Action*, report of an International Task Force convened by the World Resources Institute, The World Bank, and the United Nations Development Programme.

Index

Page numbers in *italics* refer to figures, tables and boxes

SUSTAINABLE FORESTRY HANDBOOK

Ruth Nussbaum, James Mayers, Neil Judd, Sophie Higman and Stephen Bass

'A stimulating guide to many of the issues likely to be faced in contemporary forest management.'
JAMES SOWERBY, Shell Forestry Ltd

'This book draws on the authors' considerable experience ... and the contributions of a large number of others... [It offers] comprehensive guidance ... with a large number of case-studies from around the world.'
JOURNAL OF ENVIRONMENTAL PLANNING AND MANAGEMENT

1-85383-599-4 • Paperback £25.00

THE FOREST CERTIFICATION HANDBOOK

Christopher Upton and Steve Bass

'Explains clearly how a certification programme should be run, discusses critically what certification may or may not achieve in terms of solving the problems that beset forests, and gives an up to date account of the relationship between certification and other efforts that are being made to improve forest management worldwide. I strongly recommend it to readers, whether they be forest managers, traders, government officials or simply those who are concerned for the future of forests.'
From the Foreword by DUNCAN POORE

1-85383-222-7 • Paperback £29.95

To order
www.earthscan.co.uk
Earthscan Freepost • 120 Pentonville Road, London, N1 9BR
Fax +44 (0)1903 828 802 • Email orders@lbsltd.co.uk

SELLING FOREST ENVIRONMENTAL SERVICES

Market-based Mechanisms for Conservation and Development

Stefano Pagiola, Joshua Bishop and Natasha Landell-Mills

'The success stories laid out here point to strategic directions that will carry us to a future that brings ecological, economic, and social approaches together and maintain forests in the landscape.'
From the Foreword by MICHAEL JENKINS, Executive Director, Forest Trends

'This book makes an invaluable contribution to bringing sustainable forest management one step closer.'
RICHARD MCNALLY, Economics and Global Policy, WWF-UK

1-85383-888-8 • Paperback £15.95

FOREST POLITICS

The Evolution of International Cooperation

David Humphreys

'A complete and absorbing history of a decade of intense international politics… Offers many insights for future negotiators of sustainable solutions.'
STEVE BASS, International Institute for Environment and Development

'Skillfully navigates the jungle of forest politics, leaving us in no doubt that the verbal commitment to save the world's forests has yet to be translated into action on the ground.'
FRANCIS SULLIVAN, World Wide Fund For Nature

1-85383-378-9 • Paperback £15.95

AGRICULTURAL EXPANSION AND TROPICAL DEFORESTATION

Poverty, International Trade and Land Use

Solon L Barraclough and Krishna B Ghimire

There is no clear-cut causal relationship between international trade, agricultural expansion and tropical deforestation. The primary aim of the book is to highlight the need to seek solutions in far-reaching institutional and policy reforms adapted to specific socio-economic and ecological contexts.

1-85383-665-6 • Paperback £14.95

TROPICAL DEFORESTATION

A Socio-economic Approach

CJ Jepman

'An important contribution... well written [and] well researched... this fine book will be of interest to natural scientists, social scientists and resource mangers interested in the social and environmental dimensions of the development of tropical forests, but also to scholars of international environmental economics interested in the interplay between ecological systems and human institutions.' *ENVIRONMENTAL POLITICS*

1-85383-238-3 • Paperback £18.95

TAPPING THE GREEN MARKET

Certification and Management of Non-Timber Forest Products

Patricia Shanley, Alan R Pierce, Sarah A Laird and Abraham Guillén

Tapping the Green Market explains the use and importance of market-based tools such as certification and eco-labelling for guaranteeing best management practices of NTFPs in the field. Using extensive case studies and global profiles of NTFPs, this book not only furthers our comprehension of certification processes but also broadens our understanding of NTFP management, harvesting and marketing.

1-85383-810-1 • Paperback £24.95

UNCOVERING THE HIDDEN HARVEST

Valuation Methods for Woodland and Forest Resources

Bruce M Campbell and Martin K Luckert

For conservation of plant resources to succeed, it is essential to understand their importance for local people and their livelihoods. It describes the diverse products and services provided by forests and woodlands – the hidden harvest – and sets out clearly the range of economic and other approaches to valuing them.

1-85383-809-8 • Paperback £24.95

To order
www.earthscan.co.uk
Earthscan Freepost • 120 Pentonville Road, London, N1 9BR
Fax +44 (0)1903 828 802 • Email orders@lbsltd.co.uk